# World Class Health and Safety

Getting your qualification is just the start of the safety professional's journey towards effective workplace practice. *World Class Health and Safety* doesn't repeat the whys and whats of health and safety management, instead it is a helpful how-to guide for newly qualified and experienced health and safety professionals to get the best out of their knowledge, experience and the people they work with. This book is filled with practical examples that bring the subject to life, covering the skills and techniques you need to be a leader of safety, overcome inaction and make lasting positive changes to safety performance and culture – enabling more people to go home safe every day. *World Class Health and Safety* teaches the reader to:

- work efficiently and effectively with senior managers and budget holders to implement the wider corporate social responsibility agenda;
- emphasise the 'value-added' benefits of good health and safety management clearly and simply;
- create effective and engaging training;
- use monitoring and audits to get the best out of the resources available.

*World Class Health and Safety* is essential reading for those wishing to invest in their own professional development, to communicate effectively, and to understand and deliver safety in the wider business context, wherever in the world they might be working.

**Richard Byrne** is Health and Safety Director for the Contract Merchanting Division of Travis Perkins plc and has previously worked in health and safety positions for some of the UK's largest organisations, including the Midlands Co-op, Royal Mail Group, ATS Euromaster and Network Rail. A chartered member of IOSH and associate member of IEMA, Richard is a well-known speaker, contributor to health and safety periodicals, and thought leader in the areas of safety leadership, behavioural safety and culture change.

'From strategy and culture change, through financial control and project management to the psychology of leadership and effective communication, the skills required to be successful in health and safety are both wide ranging and complex. Those with such skills will be well positioned to guide their organisations to both world class health and safety and business performance. Richard has such skills in abundance and he shares his wealth of experience in this easy to read book. It is an essential read for health and safety professionals and corporate leaders alike.'

*Gareth Llewellyn, former Executive Safety, Technical and Engineering Director, Network Rail, UK*

'What a refreshing read ... . A "must have" for the desk of any health and safety practitioner. If you are feeling ignored or overwhelmed in your health and safety role, take a break, read this book, then start afresh.'

*Rowena Jackson, BSc (Hons), CMIOSH*

'Richard's highly informative and engaging articles for SHP have helped thousands of health and safety practitioners fine-tune their approach to implementing health and safety and encouraged them to think about ways to influence and inspire their workforce.'

*Roz Sanderson, Editor, SHP Online*

# World Class Health and Safety

The professional's guide

Richard Byrne

Routledge
Taylor & Francis Group

LONDON AND NEW YORK

First published 2016
by Routledge
2 Park Square, Milton Park, Abingdon, Oxon OX14 4RN

and by Routledge
711 Third Avenue, New York, NY 10017

*Routledge is an imprint of the Taylor & Francis Group, an informa business*

*British Library Cataloguing-in-Publication Data*
A catalogue record for this book is available from the British Library

*Library of Congress Cataloging in Publication Data*
Names: Byrne, Richard (Health and safety director)
Title: World class health and safety: the professional's guide/Richard
Byrne.
Description: New York, NY: Routledge, [2016] | Includes bibliographical
references and index.
Identifiers: LCCN 2015047036| ISBN 9781138183902 (pbk: alk. paper) |
ISBN 9781315645537 (ebk)
Subjects: LCSH: Industrial safety – Vocational guidance.
Classification: LCC T55 .B97 2016 | DDC 658.3/82 – dc23
LC record available at http://lccn.loc.gov/2015047036

ISBN: 978-1-138-21499-6 (hbk)
ISBN: 978-1-138-18390-2 (pbk)
ISBN: 978-1-315-64553-7 (ebk)

Typeset in Sabon and Gill Sans
by Florence Production Ltd, Stoodleigh, Devon, UK

# Contents

# Illustrations

**Figures**

## Tables

# Acknowledgements

You would be surprised how many people have been involved in some way in making this book happen; whether it be the Shah's in London (for letting me use the office as a bolt-hole) or the people who, throughout my career, have 'given me' the situations that many of the examples in the book have been based on. To all those people – thank you.

I would, though, like to pay particular thanks to a handful of people: my family for their never-ending love, support and belief. Frank Elkins for affording me the chance to explore this opportunity, Helena Hurd, Sadé Lee and the team at Taylor & Francis for putting up with my questions. Irene Stark for her belief, foresight and help, and finally, but by no means least, Tony Higgins – a great mentor, friend and the best safety professional I have ever met.

# Foreword

As a business leader I have many priorities, one of the most important is wanting each of our colleagues to go home from work safe each day. We, as a Group, are making great progress in this arena but we still have a long way to go to getting things right. Maybe safety is an area where we should always expect to 'need to do more'.

People and organisations quite rightly place a great deal of focus on making sure that the safety and health of their colleagues, customers, contractors and suppliers is looked after. I believe that most people in business these days know what is required to keep people safe, we just need help in developing and implementing these fixes from safety professionals. I often think we underestimate how important safety professionals really are.

I've seen Richard use the leadership and management techniques he describes in this book and I continually get great feedback about the way his innovative approach is so refreshing and different to other safety professionals. Technically Richard is on the money, but the way he uses the techniques in this book to bring this knowledge to life, sets him apart and explains why he achieves the results he does in the way that he does. The techniques Richard explains in this book I'm sure will help you find those links for safety too and ultimately increase the value safety professionals can add to their organisations.

Frank Elkins
CEO, Contract Merchanting Division
Travis Perkins plc

# Introduction

When I first left school I was, for a time, an apprentice car mechanic. I clearly remember desperately wanting to buy lots of high-end tools on credit just like my colleagues at the garage did. My father, though, was not convinced and as I didn't stay as a mechanic for long, he had tremendous foresight as ever. He said to me that I should learn my trade first and eventually upgrade my tools if I wanted.

Fast forward six years and I found myself going through something similar. You see, I had graduated from a university that was, at the time, the centre of excellence for Health and Safety and I was convinced that I would 'fly' in my first job. In my head I had a brilliant qualification, so why wouldn't I? Yet it wasn't until my boss in my first proper safety role helped me take my knowledge and couple it with practical ways to 'make safety happen' that I did begin to 'fly'.

Some years later I was given the opportunity by another boss to take part in a number of leadership and senior leadership development programmes. Here I found myself learning to put theoretical knowledge about management and leadership together with practical ways to deliver amazing results, both in leading people and by applying those principles to safety.

I have often reflected on this and I don't know whether I would be where I am now in my career if it weren't for those two people. What I do know for sure though, is that, given some of the really tough challenges I have faced in my career, if I hadn't been developed by these two people I probably wouldn't have had the tools to successfully tackle those challenges.

Make no bones about it, I've been really fortunate to have had the experiences and the mentors I have, and I know that I am in the minority in the safety world. And that makes me sad. I genuinely believe that everyone should go home from work in the same state that they arrived, free from injury and harm. As safety professionals we play an incredibly important (and occasionally underplayed) role in making this happen. Unfortunately, many of us don't have the mentors, or the opportunity of the experiences that we need, to become truly effective in our roles.

My aim with this book is to try to address this imbalance by sharing with you some of the approaches I have come across that seem to consistently deliver great results. If nothing else, I hope it gives you some tools to use and perhaps spark some thoughts for you to develop your own and pass on that knowledge to others in the future.

Having experienced a number of organisations, across many industrial sectors, it never ceases to amaze me that although an organisation might make different things to another, or provide different services, in the main, the safety issues that they face are incredibly similar. That is why a safety professional's qualifications enable them to have a truly varied career. In the same way, the principles outlined in this book apply to many situations, making it highly relevant to safety professionals the world over.

This book will not tell you why you should manage safety or cover the technical aspects of what you need to do to manage it – there are plenty of other books and formal qualifications that will do that. What this book does is draw on real practical experience of ways to navigate organisational politics, apathy towards safety, perceived under-resourcing and many other frustrations of safety professionals, to help you overcome them and put effectively into practice your knowledge. All this, too, hopefully to help you deliver positive, outstanding and sustainable improving safety performance and cultural change and, of course, to be a highly effective safety professional.

Richard Byrne
Staffordshire, 2016

# Part I

# Understanding the climate

To succeed ultimately as a safety professional, whatever the role, whatever the organisation, in whatever part of the world, you need to have more tools in your kit bag than just your technical knowledge (as important as this is). In the first two chapters we will explore where health and safety fits into organisational life and society in general. It is important to understand this context and the climate we presently find ourselves in because it is from this base that future chapters build.

In very recent years in the UK there has been an acceptance that there is too much health and safety regulation which is hindering economic growth and ironically not always making workplaces safer or healthy for those it seeks to protect. Yet as the timeline in Figure P1.1 shows, over the last 25 years or so, there appears to be a correlation (although understandably lagging) between the workplace fatality rate (1) and when key pieces of legislation have been introduced.

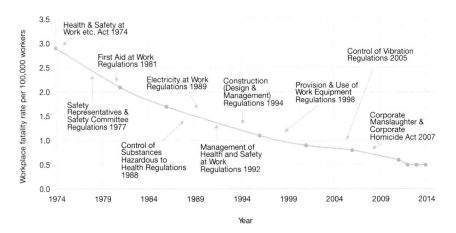

*Figure P1.1* Timeline showing UK workplace fatality rate and the introduction of key health and safety legislation (dates are approximate); taken from (1)

What this means is that there is a place for legislation and rules, yet to achieve health and safety excellence you need to move beyond these if you are ever to get anything other than simply compliance. This is nothing new. Indeed, if you consider any model of safety culture evolution, then you always start with putting controls in place like rules, laws, things that if broken can result in punishment. It is only over time do people work safely 'by choice' because of that conditioning.

Arguably, once you have your safety qualification you can 'do compliance', but to move further you need to be flexible and have different tools to use in different situations. If not, you run the very real risk of becoming stereotyped as a 'clipboard and cagoule wearing health and safety bod', which is the image the media likes to portray.

However, the world of health and safety is starting to change. It is more common for it to be talked about in the same organisational circles as other aspects of corporate social responsibility like diversity, ethical procurement and environmental management. To some in the profession this is a worry, a change to the status quo that they do not like. While there is a place for these people in the profession, like the compliance driven professionals, they stand to limit their career as a result if they are not careful.

## Reference

1   Available at: www.hse.gov.uk/statistics/history/historical-picture.pdf (accessed 14 November 2015).

# Chapter 1

# Clipboard and cagoule

If you say 'health and safety' to people these days, it conjures up a picture of unnecessary rules and a bunch of clipboard and cagoule-wearing people stopping you doing something – at least that is what the media would have you believe. In this chapter we will look at how this perception has been built up and how, at times, people in the safety profession have not helped themselves. It is only once armed with this knowledge that we can tackle this commonly held belief.

## A potted history of health and safety

The principles of health and safety or, perhaps more accurately, risk management have been with us since our ancestors were living in caves; they had to learn how to avoid or overcome wild animals that would harm them. Throughout time as well, we have seen people clad in armour in order to protect themselves from harm during times of conflict. It is only in the very recent past (the last hundred years or so) that whole body armour has become used less as combat methods have advanced.

Modern workplace health and safety has its roots firmly set in the Industrial Revolution when, after mounting pressure, the government of the day introduced prescriptive legislation in 1802 to look after the interests of children working in the cotton and wool mills throughout the UK, which was called the Health and Morales of Apprentices Act. The first factory inspectors, though, were not introduced until much later and no doubt amid much resistance from employers. Around the developed world, health and safety at work had similar birth stories.

- In the US they experienced a large number of fatalities relating to coupling and uncoupling railway units, and the application and release of railroad car brakes. Eventually, the Railroad Safety Appliances Act 1893 was introduced to provide an improved and common standard for coupling and braking mechanisms across the country.

- In 1884, Germany was the first country to introduce a worker's compensation scheme when Chancellor Otto Von Bismarck made a successful argument that people should be helped by the state if they had been injured at work.
- The province of Ontario in Canada set up a government department in 1919 to develop and enforce, among other legislation, that related to workplace health and safety.

For quite some time legislation had been prescriptive and widespread talk (let alone acceptance) of concepts like safety culture were nothing but a pipe dream, as action through compliance was the name of the game in many cases. It was not until the 1970s that the current workplace health and safety ethos came about. In America, President Nixon introduced the Occupational Safety and Health Act in 1971, which required employers to maintain a working environment free from recognised hazards and comply with various other regulations. It is still the principal piece of occupational health and safety legislation in the US.

In 1974, the UK's landmark legislation, the Health and Safety at Work etc. Act, was introduced. This, still current, legislation became more goal setting – not telling people what to do in specific terms (save for a few areas), instead outlining the principles employers and employees were expected to follow in order to achieve the goal of getting risks to safety and health to as low as reasonably practical.

This Act was a complete reversal of the existing approach and changed the way in which industry viewed health and safety. The tack was clear: 'it is your business, manage it how you see fit, but if you get it wrong the come-back will be severe.' The change from being prescriptive to being prospective is critical and was the idea of Lord Robens and the committee he chaired in 1972. They found that there was too much apathy towards health and safety and those businesses that took it seriously and engaged with the subject tended to be more prosperous than those that did it because they had to or did not do it at all.

During the early 1990s, through the European Union, all member countries moved to comply with six directives (which in the UK resulted in the '6 pack'). Among other things these introduced the absolute need to undertake formal risk assessments where work activities pose a significant risk to safety and health.

Such legislation, typically in the form of regulations, takes the Health and Safety at Work etc. Act and makes it clearer to employers what their responsibilities are towards the issue they are concerned with. For example, the Workplace (Health, Safety and Welfare) regulations add more meat to the bones of the general duties of the Health and Safety at Work Act, particularly the need to have a safe means of access and egress, facilities for welfare and the like.

The introduction of such regulations, as well as having a positive effect on reducing incidences of harm, also means that the legislation is becoming more prescriptive without 'going the whole hog' and, of course, if it is very clear what organisations need to do and they do not do it, it is easier to convince a court to convict.

Since 1974 there has really been only two pieces of health and safety legislation that have been true landmarks. We have already touched on the '6 Pack' which introduced the influential Management of Health and Safety at Work Regulations. More recently, in 2007 the Corporate Manslaughter and Corporate Homicide Act was brought in to clean up the route to prosecution where fatal accidents occurred and to make it harder for organisations to hide when things went wrong.

Since then in the UK there have been two prominent reports into health and safety legislation and its effects on business. The first was written by Lord Young (1) in 2010 and recommended a number of practical steps the UK government could take to make health and safety become more accessible and reduce the burden on business. The second, in 2011, was written by Professor Ragnor Loftstedt (2) which, while acknowledging the principles of the Health and Safety at Work etc. Act, were sound and still fit for purpose some 35 years on made a persuasive argument to readdress the balance between what the law requires and sensible risk management. Specifically he found that:

- The main problem with the legislation is the way it is interpreted and applied; there was inconsistent enforcement and third parties promoting an approach which generated an unnecessary amount of paperwork.
- Not every piece of legislation contributes, from a risk and evidence perspective, to a safe and healthy workplace.
- The UK goes beyond EU requirements for self-employed people, and they should only have to follow such approaches if they are involved in high-risk activities or can harm others in the course of their work.
- The meaning of the widely used phrase of 'so far as reasonably practical' is not well understood in small business.
- Risk assessment has turned into a bureaucratic nightmare, leading to lengthy documents being created to cover every possible risk rather than focusing on those of significance.
- The Approved Codes of Practice that accompany legislation, while viewed as helpful, are too long for employers to find the information they need.
- The EU's influence over the UK's health and safety legislation is too great, costly and not always useful, meaning that we should base our views of what becomes legislation using more objective measures of risk and evidence.

- Business feels overwhelmed by the requirements of legislation and it should be consolidated and made more industry sector specific, which would enable a 35 per cent reduction in regulations.

It does not matter which part of the world you are working in, the concepts outlined by both Young and Loftstedt are applicable to all safety professionals and the organisations they work for. The overriding message is that health and safety should make your business slicker, smarter, faster; if it doesn't, you should question whether if what you want to do is worth it.

As we have already seen, the UK has some of the best safety performance in Europe and arguably the world. While there are several reasons for this, including the UK's enforcement regime, societal expectations and the change in UK industry from the more traditional heavier industries to retail and services, one thing is clear: the need for thinking that enables business and the economy to grow while protecting people's safety and health is vitally important.

While other countries might lag behind the UK in their safety performance and expectations, organisations in the West often take their approach to safety with them when they move into markets in less developed countries and typically expect similar standards when they look to appoint organisations into their supply chain. This approach is driven mainly through the expectation of their investors and public perception and is explored in more detail in Chapter 2. This type of approach is helping the safety message to spread even when the rest of the countries' shared beliefs are not quite there yet.

## The media

It is perhaps little surprise that the newspaper press often print stories of health and safety gone mad, after all such stories have all the hallmarks of a great read. They are easy to tell and to understand, topical and involve lots of people and, possibly more importantly, they are controversial and contain an element of bad news.

The story that sparked the media frenzy about health and safety came to light in 2007 and told of a head teacher banning the traditional playground game of conkers for fear that they might shatter and damage someone's eye. If you were fortunate enough not to have read that particular story you can probably imagine how it went. But it did not stop there, ever since there has been a procession of similar stories from the banning of a pancake race through the cobbled streets of Ripon in North Yorkshire (3), to children's pantomimes not being allowed to throw sweets into the audience (4) and many more besides. In most cases you have to feel sorry for the people making these 'bad' decisions, after all presumably they acted with the best of intentions and, generally speaking, people do not want to get things wrong.

There are, of course, a number of reasons for people making these 'poor' health and safety decisions, including that in this day and age they feel as though they have to do something, otherwise they might be criticised (in other words, the problem is self-perpetuating). There is also a problem with health and safety advice insomuch as people are unable to access competent advice either because they do not know where to go for it or because they get advice from an unreliable source.

There are other reasons too, like organisations not wanting to do something and rather than saying so they look for a scapegoat – i.e. health and safety – working on the basis that it would be a brave person who argues against that, or would it? As with the Ripon pancake race, insurance companies might only offer the necessary cover with the addition of further control measures which the organisation might perceive as too onerous, and hence health and safety is cited as the reason. To an extent, you can see where the insurance companies are coming from: they are trying to reduce the chances of them having to pay out in the event of harm occurring.

All this said, the safety professional may be responsible for some of the poor decisions as well. Experience has shown that those new to the profession occasionally lack balance in their approach as they may be more focused on the textbook approach. Sadly, though, no real-life situation tends to be covered in a textbook as each business and situation will have its own unique considerations. Thus, it is important to remember that a health and safety qualification gets you 'through the door' and the textbook only provides general principles for reference; experience and continuing professional development are critical.

## Accidents still happen

Direct comparisons of accident statistics between different countries across the globe is quite difficult, as while most use standard definitions there are issues to do with data reliability. The most reliable measure, though, is the workplace fatality rate, which in the UK and the rest of Europe (5), US (6) and Australia (7) over recent years is showing a good declining trend. For less developed countries, this may not always true.

With these statistics it is tempting, particularly in times of economic decline, to take your foot off the accelerator when it comes to improving safety. Yet recent high-profile incidents have shown just how easy it is for things to go wrong. All of these dreadful incidents go to show one thing: health and safety at work and the role of safety professionals is extremely important.

### Texas oil refinery explosion

In March 2005 there was a release of heavier than air hydrocarbon vapour from a liquid overflow at a Texan oil refinery in the US. It came into contact with an ignition source, believed to be a running engine. The explosion and fire that followed claimed the lives of 15 workers and injured 180. Various issues where identified in the investigation, including cost-cutting, a lack of investment in the plant, too much focus on occupational safety instead of balancing it with process safety, a 'tick-box' mentality to checking off safety policy requirements even when they have not been met, and a poor safety concern reporting culture (8).

### Buncefield fire and explosion

Thankfully, nobody was killed, although 43 were injured, when there was a series of explosions, followed by a fire, at the 5th largest oil storage terminal in the UK on 11 December 2005. The fire burned for five days and was the largest fire seen during peacetime in the UK. The immediate cause of the incident was that a high-level switch failed to operate and its monitoring system failed to detect the fault (9).

### Deepwater Horizon explosion

The oil rig exploded on 20 April 2010 off the Louisiana coast in the US, killing 11 workers and injuring 17 as a direct result of a well blow-out during a temporary well abandonment. A range of latent failures have been identified, including a culture focused too much on personal safety rather than major accident prevention and inadequate management systems (10).

### Rana Plaza factory collapse

This incident happened in April 2013 in Bangladesh and involved an eight-storey clothing factory collapsing, killing more than 1,100 people. The report into the incident suggests that the building should never have been built on the swampy ground, it was built of poor quality construction materials and, despite cracks appearing in the walls the day before the incident, workers at the factory were made to go to work all the same (11).

## Conclusion

The principles of health and safety at work are firmly based on sensible risk management and objective evidence that backs up the decisions we make. While some people and organisations continue to make poor safety decisions, which in turn give the profession a bad name, we too as safety professionals need to examine what we do and how we do it. The real crooks of a modern safety professional's role is to put in place a programme that develops a positive safety culture; one where everyone is engaged and aligned to working safely. Often in our keenness to do this, we can go a little over-board and have the reverse effect. To be a truly effective safety professional in the twenty-first century, you cannot rely on your technical knowledge alone to get you through. You need to be balanced, inspirational, encouraging, supportive and be able to adapt quickly.

## References

1  www.gov.uk/government/publications/common-sense-common-safety-a-report-by-lord-young-of-graffham (accessed 14 November 2015).
2  www.dwp.gov.uk/policy/health-and-safety (accessed 14 November 2015).
3  www.telegraph.co.uk/news/uknews/1577651/Health-and-safety-rules-trip-up-pancake-race.html (accessed 14 November 2015).
4  www.dailymail.co.uk/news/article-1341615/Panto-killjoys-ban-stars-throwing-sweets-children–outlaw-genies-pyrotechnic-entrance.html (accessed 14 November 2015).
5  www.hse.gov.uk/statistics/european/ (accessed 14 November 2015).
6  http://onlinelibrary.wiley.com/doi/10.1002/ajim.22258/abstract (accessed 14 November 2015).
7  www.safeworkaustralia.gov.au/sites/SWA/about/Publications/Documents/910/key-whs-stat-2015.pdf (accessed 14 November 2015).
8  www.csb.gov/assets/1/19/CSBFinalReportBP.pdf (accessed 14 November 2015).
9  www.hse.gov.uk/comah/buncefield/buncefield-report.pdf (accessed 14 November 2015).
10  www.csb.gov/assets/1/7/Overview_-_Final.pdf (accessed 14 November 2015).
11  www.theguardian.com/world/2013/may/23/bangladesh-factory-collapse-rana-plaza (accessed 14 November 2015).

# Chapter 2

# Corporate responsibility, safety and business activities

Health and safety is part of the much broader Corporate Responsibility (CR) agenda and in this chapter we will explore how this is becoming a big driver for change in many organisations, as well as working through some of the opportunities this presents for safety professionals. We will also look at where health and safety fits into organisational life, because understanding this will enable safety professionals to have a smoother ride than those who think that just because they have the law on their side, that is enough to affect change.

## Corporate responsibility

This is an area of organisational life that is called different things by different people: corporate social responsibility, corporate responsibility, sustainable business, to name a few. Whatever the name, the meaning is the same: it is about doing the right thing.

Companies have been doing the right thing by their shareholders for years: delivering profit and in turn good dividends, but now it is much more than that. How you get there is important as well. In other words, do you do the right thing by all your stakeholders: employees, suppliers, customers and the communities the organisation operates in, both in your country and beyond? Figure 2.1 outlines the key components that make up the CR agenda.

As a concept, CR goes back to the 1950s with organisations doing good deeds for society. However, it was not until the 1990s that it became more accepted and in the 2000s it was recognised as a strategically important issue for business (1).

Years ago, this might have been seen as philanthropy, where a small minority of people tried to enhance the lives of others by so-called 'do gooding'. Nowadays, organisations are realising that by developing their thinking and approach to become more of a responsible business, giving something back to the communities they operate in and going well beyond legal requirements, really does have bottom-line benefits such as:

- increased employee retention and engagement;
- improved risk management in supply chains;
- achieving competitive advantage.

The ideals of CR are considered so important that they were enshrined in law with the introduction in the UK of the Companies Act 2006 (2). This places general duties on company directors, among other things, to:

- consider the interests of their employees in business decision making;
- develop business relationships with suppliers and customers;
- consider the impact of their operations on the environment and the community they operate in;
- act to maintain a reputation for high standards of business conduct.

The Act also requires that quoted companies, through their Directors' Annual Report, provide a review of the how the business is managing its obligations to its employees, the environment and the community. These provisions, which are almost hidden in the Act, form the basis for the CR agenda.

In the UK and other developed nations, legislation exists to cover the majority of the components that make up the CR agenda. In developing and emerging economies, such provisions sadly are not always available. Forward-thinking organisations do not just rest with putting 'a tick in the box' that says 'we comply with the relevant legislation' they influence their supply chain to do the same, as well as looking for the next piece of best practice to stretch their performance.

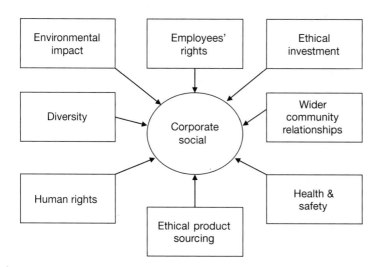

*Figure 2.1* Key components of corporate responsibility

Demonstrating that an organisation is 'ethical' is important because it is fast becoming what the public and investors expect. In the previous chapter, the power of the media was discussed and, with this in mind, it is clear how embracing CR is not only the right thing to do for the people in the organisation but also to help deliver competitive advantage; or, put another way, to help the company stand out in the crowd.

## Synergies

There are many similarities between health and safety and the other topic areas that make up CR – arguably those synergies that are strongest with environmental management.

### Environmental management

The link between health and safety and the environment is one that many individuals and organisations exploit – and why not with the amount of similarities between them?

#### Legislation

There are clear examples of when the two topics combine to form 'meaty' legislation like the Control of Major Accident Hazard Regulations (COMAH). Among other things, these regulations require the provision of adequate emergency planning, where the health and safety aspect is interested in making sure that people both on and off site are safe, and the environmental aspect considers the protection of the local and global environment should an incident occur.

That said, health and safety legislation tends to be based on a risk-management approach, while environmental law centres upon the use of best available options and goal setting. However, do not be fooled because both areas have all-encompassing acts – the Health & Safety at Work etc. Act places the general duty on an employer to provide a safe working environment, which arguably they cannot have done if an accident occurs. Similarly, the Environmental Protection Act 1990 places a 'duty of care' on employers, so should an environmental incident occur, the company could still be prosecuted under this duty despite the fact that they fell outside a particular set of related environmental regulations.

#### Management systems

Both topics rely heavily on the use of management systems and, as we will discuss in a future chapter, management systems are management systems. Some of the names might change, but essentially all they are trying to do is manage an issue in a structured and systematic fashion. So similar are they that options are also available to integrate environmental and health and safety management systems to make a 'one-stop shop'.

### Tools: risk assessment

There are striking similarities between a health and safety risk assessment and the environmental aspect and impact process. Environmental aspect takes the work activity, project or product and asks what could happen, while the environmental impact asks if that can happen, what impact would it have on the environment. The end result is to focus on ways to mitigate any negative effects. Putting hazard and risk together in the form of a risk assessment is much the same thing as thinking about what could cause harm to people as a result of work activities.

### Tools: safe systems of work

Equally, environmental operating procedures are similar to safe systems of work; both explain how to do the job to prevent harm occurring either to the environment or people.

### Tools: accident reporting and investigation

Just as accident reporting is vital to a safety professional measuring the success of the overall safety system and identifying areas for improvement through investigations, so too are incident reports and investigation of environmental breaches or issues for the environmentalist.

### Driving efficiencies

Good health and safety management not only reduces accidents but should also improve productivity and efficiency, and thus be margin enhancing or cost reducing. Well thought out environmental programmes like that of recycling a firm's waste through segregation of waste streams, having larger waste receptacles and reduced collections can become cost neutral or even revenue generating.

### Engagement

At the heart of all of this, though, is the need to get everyone within the organisation truly engaged. Whether that is getting the directors' attention or requiring employees to follow the procedures that have been put in place, engagement is key to health, safety and environmental management.

## Other topics

It may not be immediately obvious where else the safety professional can influence or positively contribute to the rest of the CR agenda, as traditionally health and safety has stopped at the factory gates, dealing as it does with the way work is managed inside those confines. There are opportunities there but you have to open yourself up to more contemporary ways of thinking in order to realise them.

### Human rights, ethical product sourcing and investment

For an organisation's approach to CR to be sound and, for the necessary assurances to be provided, there is an obvious need for the people from whom they buy materials, products and services to be working towards the same CR objectives. If they are not, it makes a mockery of what the organisation is trying to achieve.

Most organisations will ask for information on health and safety performance and arrangements, but are they asking the right questions and does anyone ever check the answers they get back? There is a real opportunity for the safety professional to add value to this process by not only making sure sensible information is requested – that is, sensible across country and continental boundaries – interpreted in the right way and ultimately independently verified. This is a logical extension to the arrangements organisations have to vet contractors who undertake work for them.

The same can be applied to investments and, while admittedly they use different mechanisms, would you want your pension fund invested in a business that was killing or seriously injuring people every week because they did not have suitable safety arrangements in place?

### Supporting the wider community

Health and safety can add to delivering the CR agenda by supporting the wider community through sponsorship and awareness, and if used in the correct way they can help to achieve the broader strategic aims of the safety profession. Many sponsorship opportunities are available with safety-related links that go beyond workplace safety and enter into road and home safety:

- sponsorship of 'safety event' – e.g. a safety award in suppliers' or customers' industry sector;
- sponsorship of student – e.g. NEBOSH award, undergraduate placement, M.Sc., Ph.D.;
- donation to safety-related charity – e.g. RoSPA, local safety group, BRAKE;
- delivery of health and safety awareness training in schools and colleges;
- provision of health and safety support to local community groups.

## Where health and safety fits into organisational life

Health and safety is not the sole purpose why businesses and organisations exist. In the main, their sole purpose is make a profit or provide a service, and they should be able to do that without hurting people or adversely affecting their health. In much the same way, they should pay fairly and not discriminate against people. Often people say that health and safety is the most important thing in an organisation, their number one priority. This works to a point, but in a world of competing priorities people soon start to lose faith in this rhetoric. The reality is that health and safety should be one of the organisation's top priorities with a healthy tension between other areas of the business like sales, operations, production. Here are two high-level examples to bring this concept to life:

- if an organisation has too much focus on sales and insufficient focus on safety, the chances are at some point there will be an incident that will damage the organisation;
- too much focus on safety and not enough on sales will ultimately lead to an organisation that is unsuccessful.

The safety professional's role, in many cases, is to help create that healthy tension in whatever setting they find themselves, from the board-room to a front-line team briefing. How you do this is key, though, as done the wrong way you will end up being seen as the 'clipboard and cagoule' wearing stereotype – we will explore in later chapters how you can do this important aspect of the role right.

### Clues as to how an organisation views safety

Experience has shown that you can often learn a great deal about an organisation's view of health and safety (and the type of role their safety professionals do) by the route in which they report into the Board or senior management team – for example:

- safety functions reporting in through legal, audit or property functions tend to be more compliance driven;
- teams reporting through the operational line can, at times, end up being the safety doers, rather than the enablers of safety change;
- those that report through HR tend to be more focused on cultural change, with those reporting to the CEO or MD being more balanced and drivers of change.

None of these are wrong – far from it – provided that they reflect where the organisation is on its safety maturity journey.

Likewise, experience shows that job titles also give some indication of how the organisation views safety and the role of the safety professional. The title 'Safety Officer' tends to be used where policing of rules is more necessary. If you look across high hazard or safety critical industries (who are generally viewed as leading the way for safety performance, innovation and culture), they tend to have a director or head of role that sits on the management board.

### Links to other organisational functions

While the immediate community of health and safety management tends to be confined to safety and occupational health professionals as well as trade union safety representatives there are, of course, many other very influential stakeholders each with a different reason for being interested. Some of these internal stakeholders and their areas of potential interest between their spheres of work and that of health and safety are shown in Table 2.1.

Don't forget that there are external stakeholders too who have similar, or in some cases more, influence than those within the organisation – for example:

* shareholders;
* enforcing authorities;

*Table 2.1* Internal organisational stakeholders and their areas of potential interest between their spheres of work and that of the safety professional

| Stakeholder | Areas of interest in health and safety topics |
| --- | --- |
| Human resources | Terms and conditions of employment |
| | Employee retention |
| | Sickness absence |
| | Disability discrimination issues |
| | Stress at work |
| | Engagement |
| | Employee and management development |
| | Drug and alcohol misuse programmes |
| Finance | Controlling costs |
| | Return on investment |
| Sales and marketing | Competitive advantage |
| | Marketing tool |
| Property | Resolution of estate-related issues |
| Operations | Reducing lost work days due to accidents |
| | Well-defined workable processes and procedures |
| | Solutions 'fit for purpose' and proportionate to risk |
| Employees | Consultation |
| | Engagement |
| | Development |
| | Processes and procedures that do not slow the job |

- customers and suppliers;
- insurers;
- trade associations.

## Conclusion

The world of work is changing, and health and safety is starting to get swept up into the bigger issue of CR, which is no bad thing as it enables health and safety to be raised even further on the corporate agenda. The safety professional should not shy away from, or resent the 'interference' of, CR as it provides significant opportunities for health and safety to shine both through the provision of the competitive edge and through the use of safety professionals to support the wider programme. Our training is more than just technical knowledge; it is also about management principles which can be applied to almost any subject.

We also need to remember that we are part of an organisation and there are others in the organisation who believe their role is vital too. The truth, of course, is that all the roles in an organisation are important and we need to understand that while the role we do is hugely necessary, we need to link into other professionals and colleague populations to help us enable a successful organisation.

## References

1   www.emeraldinsight.com/doi/abs/10.1108/1747111111117511 (accessed 14 November 2015).
2   www.legislation.gov.uk/ukpga/2006/46/pdfs/ukpga_20060046_en.pdf (accessed 14 November 2015).

# Part 2

# Leadership and management skills for safety professionals

If the 'clipboard and cagoule' approach is one that the profession needs to get away from, the obvious question is 'What does a world-class safety professional look like?' In short, the role that they perform is quite different from the traditional approach, as Figure P2.1 shows, it is probably only about 20 per cent 'technical' health and safety and 80 per cent leadership and management.

Traditionally, health and safety has been the job that you get given when the organisation wants you 'out of the way' or you are on the 'wind down' to retirement. Now, though, it is used as a stepping stone into more senior management positions because at the end of the day being a world-class safety professional is about good leadership and management, and every organisation needs these skills.

The role is not about banging your fist on the table and demanding that people do things; it is about:

- providing accurate information about safety issues facing the organisation and the likely impact (both in terms of safety and financial performance);

**80% Leadership and management skills:**

- Strategy and planning
- Communicating effectively
- Improving policies and processes
- Developing business cases
- Report writing and presentations
- Managing conflict
- Problem solving
- Leading and managing change
- Project management
- Time management
- Coaching
- Overcoming inaction

**20% 'Technical' health & safety:**

- Risk assessment
- Accident investigation
- Auditing
- Accident statistics

*Figure P2.1* Make-up of the modern safety professional's job role

- providing a range of workable and cost effective solutions (with a preferred option) to the issues noted;
- enabling cultural and behavioural change.

The idea is that the organisation's leaders can use the information to make informed business decisions about how the organisation is going to manage its risks. In other words, the safety professional provides the advice and guidance, but ultimately the decision is taken in a broader business context, which is the right thing to do.

To do this, it should be obvious that you need a degree of underpinning technical knowledge. There are many different educational routes to achieve this, depending on your preferred learning style. There are a number of professional bodies which you can become a member of, with your membership level depending on the level of qualification and experience you have.

In the following chapters, though, we will explore each of the leadership and management skills in Figure P2.1 and show how they apply to the safety professional's role (the only one we will not cover is overcoming inaction as that particular subject is the focus of Part 3). It is these skills that will really help you to become even more effective in your role. Indeed, as we explore each of these skills, you will see that they all have at their heart a similar approach.

# Chapter 3

# Creating a safety strategy and work plan

Often it is easy to get too involved with the detail of things, which occasionally makes it difficult to see the wood from the trees. Every function within an organisation should have a clearly defined strategy linked to the overall aims and objectives of the organisation. Getting this bird's-eye view can sometimes be a challenge either because life is too fast paced or because we do not have the tools to do so.

The first step towards improving any organisation's safety performance and culture is to work out a strategy and a plan of how you are going to get to the end result. The safety professional's role in this should be obvious, yet from experience this tends to be something that is not covered by any formal safety qualifications, nor is it something that safety professionals do very well. It should be far more than just what, how, who and when.

In this chapter we will explore how you can develop an effective strategy and work plan using principles that apply equally to someone working on a single site, in a geographical area or setting the strategic direction for a global organisation.

## What is a strategy and why do you need one?

A strategy is simply a plan to help you achieve a long-term aim. Without one, particularly in safety, you can find yourself lurching from one thing to another and often find yourself not really getting anywhere. Not only can having a clear strategy help you overcome this, but it can also be a useful reference point to go back to when you are being challenged as to why you are doing something and, of course, you can challenge yourself to see if what you are doing is helping achieve the overall objective.

Developing a strategy is based around three basic questions:

- Where are you now?
- Where do you want to get to?
- How will you get there?

There are numerous ways to develop a strategy; simply type it into an Internet search engine and you will be amazed at the number of results. There is one tool, though, that experience shows is versatile yet easy to use and actually comes from the world of marketing. It is called SOSTAC® and is the registered trademark of PR Smith (1) and it is adapted here, with kind permission.

## SOSTAC®

The model was created during the 1990s by PR Smith (www.PRSmith.org/SOSTAC), a marketing expert based in London (see References at the end of this chapter for more details) and while focused on developing a marketing strategy, experience shows that it works as well as a framework to develop a safety strategy, just as it does for other business-related activities. The name stands for:

- Situation Analysis – where are we now
- Objectives – where do we want to go
- Strategies – summary of how we are going to get there
- Tactics – outlines the detail of strategy
- Actions – who is going to do what, when and how
- Control – how will we make sure we are going in the right direction.

The model takes you through a structured thought process to make sure you are aiming for the right things, pulling the most appropriate levers to achieve them and helping you consider how you will check that you are on track.

### Situation analysis

The key to developing a successful strategy is to make sure you make the right decisions; you can only really do this if you *really* understand where you are now. To do this for an effective safety strategy you can refer to various data sources such as accident statistics, insurance claims and audit results (internal or external). However, data only really tells one side of the story.

To make sure you have a rich picture of what is going on, you need to 'triangulate' the data by physically going to see for yourself what is happening as well as listening to what others (colleagues, regulatory bodies) are saying, as Figure 3.1 shows. We will return to this idea in a future chapter.

It is only by bringing all three elements together that you will get a reliable and valid picture. This is then normally presented in a SWOT analysis:

- Strengths – what are we doing now that is good?
- Weaknesses – what are we doing now that needs improving?

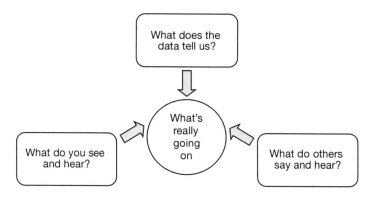

*Figure 3.1* Triangulation helps ensure that you have all the information necessary to make informed strategic decisions

- Opportunities – what could we do differently in the future?
- Threats – what could happen in the future that could compromise us?

## Objectives

Here we explain where we want to get to. These aims can be a mixture of soft or hard measures depending on the subject but they must be SMART:

- Specific – exactly what is it you want to do?
- Measurable – how will you measure success?
- Achievable – can you really achieve it?
- Realistic – is it too difficult?
- Timely – is there a deadline?

## Strategy

This is about listing the general approaches you are going to take to achieve the objectives you have set – in other words, what are the headlines – e.g. if you want to improve your culture by driving more ownership and colleague engagement you might need to streamline your processes, hence the strategy here might be 'simplification'.

## Tactics

Based on the strategy you have outlined, in this section you detail the most effective tools or ways to achieve the objectives. In short, you are fine-tuning your plan of attack. The trouble is when you are brain-storming ideas, how do you know which are the ones that will have the most impact?

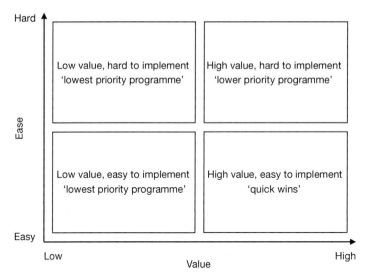

*Figure 3.2* A commonly used programme prioritisation tool

A useful way to help overcome this is to use a prioritisation tool. As an initial filter if the thing you think you should do is related to achieving legal compliance in an area which currently poses a threat to the organisation, this should be a much higher priority over others.

Such prioritisation tools can be complicated but the following is a simple tool that provides a structured and justifiable methodology which is used by many organisations. The tool, shown in Figure 3.2, is based on two measures:

- ease of the programme's implementation (in terms of time, cost and effort);
- amount of value the programme has in relation to achieving the strategy.

Another interesting consideration in determining the priority in which things should be implemented is that of the organisation's 'appetite' for it. If the programme is an added value item and the organisation is not ready for it, why expend all that time, money and effort on it when it could be used more effectively elsewhere? The key is to work out where you will get the biggest bang for your buck.

### Actions

This is the simple action plan with which most people are familiar. It is here that you decide who will do what, when and how.

## Control

This section is often overlooked when developing a strategy or plan, and it is about how you make sure you monitor the implementation of the plan to achieve the strategic outcomes you have committed to. It is sometimes known as governance.

## Putting it into practice

Having discussed the theory, we will now use two situations to show how it translates into the work of a safety professional.

### Case study 1: new organisational safety strategy

A safety professional has recently joined a new organisation and has been tasked to develop a new safety strategy for the next five years.

*Table 3.1* Situation analysis

| Data reviewed | What you have seen | What you have been told |
| --- | --- | --- |
| Employee engagement scores<br>Accident reporting rates<br>Audit trends<br>Employee retention rates | Accidents and near misses going undetected/reported.<br>Detailed safe systems of work that are unnecessarily onerous.<br>Generally people are positive about the firm. | People don't report incidents because they think they will get blamed for them.<br>Too many rules in place that hinder working efficiently and safely.<br>People quite like working here, there are worse places. |

*Table 3.2* SWOT

| Strengths | Weaknesses |
| --- | --- |
| Employee commitment to the firm.<br>Understanding that working safely is important.<br>Management desire to improve things.<br>Well-defined basic processes and good levels of legal compliance.<br>Generally improving accident frequency and severity rates. | Some safety rules are too onerous for the risk they are designed to manage.<br>Perceived blame culture among front-line employees and managers.<br>Poor reporting rates. |

| Opportunities | Threats |
| --- | --- |
| Readdress the balance between risk and control.<br>Improve employee ownership and involvement.<br>Celebrate safety success and promote things done 'right'. | No horizon scanning is taking place, meaning we don't know what changes are coming that we need to take account of in our thinking and budgeting process. |

*Table 3.3* Objectives

|  | What will it be like to work here? | We will feel good when people tell us about incidents and near misses.<br>Safety will drive us to work more efficiently.<br>Everyone truly believes in the value safety brings to our business |
|---|---|---|
| By the end of the five-year cultural improvement strategy: | What will we achieve? | 50% reduction in reportable accidents.<br>100 lost work days saved.<br>15-fold improvement in near-miss reports.<br>75% reduction in audit non-compliances. |

*Table 3.4* Strategy

We will improve our safety culture by starting a behavioural-based safety programme drawing on:

- simplified processes; and
- clear leadership support.

*Table 3.5* Tactics

We will:

- Use front-line colleague risk-assessment groups to streamline our processes.
- Introduce and use health and safety forums to engage employees with making safety decisions.
- Make incident reporting easier.
- Promote a safety recognition programme.
- Develop our leaders to effectively lead safety, and so on.

*Table 3.6* Actions

| Title | What | Lead | When | Cost |
|---|---|---|---|---|
| RA Groups | Set up frontline employee risk assessment groups | T. Close | June | Nil |
|  | Provide risk assessment training to groups | N. Davis | July | Nil |
|  | Determine list of assessments to be reviewed and in what order they need to rewritten | T. Close | July | Nil |
|  | Working groups to review assessments and make recommendations to the site health and safety forum | RA Groups | May Next Year | Nil |

Table 3.7 Control

| What? | Where? | How? | Who? | Frequency |
|---|---|---|---|---|
| Progress will be monitored | Board meeting | Agenda item | Functional director responsible for safety | Quarterly |
| Changes to the plan | Site management meetings | Report and agenda item | Safety professional | Monthly |
|  | Board meeting | Agenda item | Functional director responsible for safety | As necessary |

## Case study 2: noise reduction strategy

A safety professional has been tasked with developing a noise reduction strategy in the machine shop of a manufacturing facility.

Table 3.8 Situation analysis

| Data reviewed | What you have seen | What you have heard |
|---|---|---|
| Noise assessment results. Insurance claims for noise-induced hearing loss. Existing safe systems of work. Maintenance records for machines and equipment. Health surveillance arrangements. | Poor maintenance of machines – e.g. loose panels. Employees not always wearing hearing protection. Managers never wearing hearing protection. | 'Nobody listens to us.' |

Table 3.9 SWOT

| Strengths | Weaknesses |
|---|---|
| Existing noise assessments are available but were carried out some years ago. Processes and budget exists for machine and equipment maintenance. | Processes for planned preventative maintenance not followed. Inadequate signage for hearing protection zones. Little compliance with the need to wear hearing protection. Health surveillance provision is poor. |

continued . . .

*Table 3.9* Continued

| Opportunities | Threats |
| --- | --- |
| Over-reliance on hearing protection rather than trying to remove or reduce the noise at source.<br>Education programme would greatly help improve compliance and engagement.<br>Employee safety forums could address issues being highlighted and play to wider improvement programme.<br>Productivity increase through reduced machine downtime due to breakdowns. | Continued losses will occur from employers' liability claims.<br>Potential for enforcement action if employees frustrations grow. |

*Table 3.10* Objectives

| | |
| --- | --- |
| By the end of year 3 we will have: | Suitable and sufficient noise assessments.<br>A replacement programme for machines that are too noisy. (*)<br>Effective planned preventative maintenance of all machines and equipment.<br>Achieved a 20 per cent reduction in machine and equipment down time as a result of improved poor maintenance<br>Everyone protecting their hearing.<br>25 per cent improvement in health surveillance scores in relation to hearing. |

(*) criteria of what is too noisy will be determined once noise assessments have been completed

*Table 3.11* Strategy

We will reduce the amount of noise collegues are exposed to through a safety improvement programme based on:

* accurate risk information;
* capital investment;
* machinery maintenance; and
* collegue education.

*Table 3.12* Tactics

We will:

* use external experts to accurately measure the noise created by our machines;
* develop a machine replacement programme based over a three-year horizon;
* create a business-wide education programme about noise and hearing;
* strengthen our health-surveillance programme;
* link into our safety leadership programme and lead by example.

*Table 3.13* Actions

| Title | What | Lead | When | Cost |
|---|---|---|---|---|
| New Kit | Determine exact costs of poor noise management | D. Fox | 30.10.16 | Nil |
| | Develop business case to invest in new machinery based on the predicted 'health' costs of not acting and the down time from old machines | A. Wills | 30.11.16 | Nil |
| | Present business case to management for sign off | B. Jones | 06.01.17 | Nil |
| | Establish roll out programme steering group and start to deliver new machines | R. Groves | 01.04.17 | TBC |

*Table 3.14* Control

| What | Where | How | Who | Frequency |
|---|---|---|---|---|
| Progress will be monitored. | Health and safety committee meeting | Presentation | Safety professional | Monthly |

## Conclusion

In this chapter we have seen how you can use approaches from other areas of business to help add structure and logical thinking to safety. Taking this sort of approach to safety, experience shows, is incredibly powerful and means that you are more likely to succeed in delivering not only the matter in hand, but be able to take on more things and be successful at delivering those too. To deliver any strategy it is also important to determine the cost and benefits of doing so. This is covered in detail in Chapter 8.

In a world of competing priorities and limited resources, strategic thinking is a fundamental skill and essential for the modern safety professional.

## References

1   SOSTAC® is a registered trade mark of PR Smith. SOSTAC® videos, infographics and books from: www.PRSmith.org/SOSTAC
Smith PR (2013) The SOSTAC® Guide To Your Perfect Marketing Plan, www.PRSmith.org/SOSTAC
Smith, PR (2015) The SOSTAC® Guide To Your Perfect Digital Marketing Plan, www.PRSmith.org/SOSTAC
Follow PR Smith's daily updates: *t: PR_Smith f: PRSmithMarketing*
SOSTAC® Certification from www.PRSmith.org/SOSTAC

# Chapter 4

# Communicating safety

Communicating any sort of information is a critical part of most people's jobs and it is particularly important to safety professionals. Yet from experience you would be surprised how many people and organisations make a hash of it, which is a shame if you think about it, because all communication is, at the most basic level, one person sending a message to another, who interprets it and responds accordingly.

Despite the fact that in the twenty-first century there are far more communication tools available to us, it is sad that our standard response to communicating safety messages to people is either through posters, newsletters or email. While these particular tools have a place, we are potentially missing opportunities all the time to get a safety message across by having a limited toolkit to use. Not only that, but we are not very good at getting multiple communication flows going – in other words, information does not always flow up, down and side to side as it perhaps should.

In this chapter we will use the formula shown in Figure 4.1 that experience shows can help overcome these challenges. In doing so, we can improve safety culture and performance far beyond just making sure we 'tick the compliance box' to do with the legal requirements of providing employees with information and instruction.

It is worth saying that we will approach it from how we communicate information to colleagues rather than how they communicate to us. This said, though, for each of the tools we discuss, it will become apparent how they can be used (or not, as the case might be) to encourage that two-way communication.

---

**Effective Communication** = (Clear Message + Right Communication Mediums) × Repeated Exposure

---

*Figure 4.1* A formula for effective communications

## A clear message

It might sound obvious, but the first step to achieving effective communication of safety is to work out what you want to say. If you think about it, every organisation wants to get lots of different messages across to its people. There is nothing wrong with this. It is life in a modern workplace after all, but at any one time there are generally too many messages bouncing around. This means that people either:

- don't get the message(s);
- get part of the message(s);
- tune into those messages to do with the thing they are interested in.

A safety professional did a snap survey of the communications being 'pushed' out to colleagues in their workplace. They found that there were messages, about the:

- business's current trading performance and the steps needed to improve this;
- new contracts won in the previous month and how they are to be serviced;
- next month's marketing campaign and how this would affect colleagues' bonus;
- the launch of the upcoming employee engagement survey;
- new product recalls.

As well as the monthly safety tool-box talk topic and all the local issues, the management team needed to get across to their team.

As the example above illustrates, no matter what you are trying to communicate, you are always competing against other messages. Therefore, it is crucial that, from all the things you want to get across, you pick the most important one. To do this, make a list of the things you want to talk about and then be really critical about that list to get to *the* message you are going to send (Chapter 11 on problem solving has some techniques that might help with doing this).

## Using the right communication medium

Having decided about the message you are going to 'send', you now need to think about the next part of the formula: the way you get the message across. This is an area that breaks down into two fundamental questions:

- who is the audience you are trying to communicate with?
- how do they like to be communicated with?

Often these questions are not asked when people are planning their communications; instead, they think about the best way for them (the sender) to get the message out there.

### Who is your audience?

This should be an easy question to answer – for example, do you want to communicate:

- to the whole organisation;
- the management population;
- a specific group of front-line colleagues;
- several different groups of people;
- colleagues who tend to do manual work and so have no work or little computer access;
- your mobile sales force or people who work from home?

Understanding this is really important because each group of people will prefer being communicated with in different ways to the next group; if you can identify this and, then make sure you do it, you stand a far better chance of getting your message across and for it to be remembered.

### What is the best communication medium?

Having identified who you want to get your message across to, you can think about how they like to be communicated with. You could, if you wanted to and had the time, survey all your workers to find this out. However, for most of us this is unnecessary as generally experience shows that:

- 16–24-year-olds prefer to be communicated with via social media and are more open to communication technology, although this is spreading to older people as well, as tools like Facebook and Twitter are becoming part and parcel of modern life.
- Front-line employees in manual-type roles tend not to like written communication as they do not always have the time to read it and it is not written in an easily accessible way; similarly, they have limited access to work computers and intranets.
- Front-line managers, while accepting emailing and communications on the organisation's intranet, tend to prefer different ways of being communicated with, as messages can often get lost in the sea of emails they receive on a daily basis.
- Middle and senior managers are comfortable with written communication and are more likely to read it if comes via an email.

These, of course, are just rough rules of thumb and to develop a richer picture of how people in your organisation like being communicated with, you could as part of your planning go along to your safety forum (see Chapter 21) and ask for their help in suggesting the right method.

### Repeated exposure

Having a clear message conveyed in a way that your target audience prefers will help increase the likelihood of people picking up the message and remembering it. But you cannot leave this to chance. You need to repeatedly expose them to that message in whatever way they want to receive it. In other words, pressing 'send' on an email and thinking people will act on what you say is somewhat naive. Just like the famous Russian scientist, Ivan Pavlov, who in the early 1900s conditioned dogs to salivate when a bell was rung through repeated exposure to meat every time it sounded, you need to keep repeating the message to help it sink in and elicit a correct response. The difficultly here is that if you over-expose people to the message they get annoyed and they lose interest, so experience would suggest repeating the message about three times is enough for the message to start to sink in for most people.

### Communication tools

Having gone through the formula, we now need to expand the range of communication tools we have in our tool box. What follows are a number of different ways to communicate, some of which will be very familiar and some of which won't be.

### Posters

These are widely used to promote messages from information about sales promotions and charity events, as well as reminding colleagues to follow safety rules. When they are first put up, they are at their most impactful, although this is still a relatively poor way to effectively communicate with people. Inevitably, they will always be part of communicating a message – but they are just that, part of the solution. To make posters effective, they need to be eye-catching, snappy, easily understood and replaced regularly.

### Emails

Electronic mail has, in the majority of cases, replaced memos and, increasingly, letters. The fact that they are easier to send means that people send them all the time, but that also means that on any given day people have lots to read. Just because you send an email, particularly en masse, there are few ways to ensure that people actually do read it, yet they are good for getting a message (at least the headlines) out to the masses in a quick and efficient manner.

Making emails effective means that the language needs to be appropriate to the audience, snappy and, when open, should not take up more than the screen. Experience suggests that when people have to scroll down or if reading the message takes time, they tend to lose interest fast. Thinking who the actual email comes from is also useful: are you more or less likely to read an email that has come from the managing director? Of course, you cannot do this for all your communications – hence the need to pick the right tools for the job. Where possible, it is also worth refraining from asking people to forward or cascade the mail throughout their team as invariably all they do is forward it with 'FYI' rather than putting it in the right context with their own personal stamp on it.

### Intranet

Organisations tend to spend reasonable sums of money in developing and maintaining their intranet site and there is a tendency to use this as one of the main sources of communication. Using an intranet effectively for safety communications should only be used to back up other communication tools unless your organisation has a really good hit rate on their site, as access and use is surprisingly limited.

### Enclosures

Arguably one of the most traditional ways to communicate your message is to include text referring to it on payslips or sending a leaflet out with them, the logic being that everyone looks at their pay slip, so you have more chance of at least getting them to see the message; of course, that does not mean that they will register it.

### Letters home

As emails and online banking take over as the norm, the amount of 'meaningful' post delivered to our home addresses is becoming less and less. This presents an opportunity to us, as for the right message people are more likely to open a letter and read it; not only that, but you might just generate a conversation between your employee and their family about the letter and what it was about. As with using email, you should only use letters sparingly and as a general rule of thumb they should cover no more than one page.

### Displays

Although these do not work so well for multi-site organisations, having a display in a prominent area with high foot-fall in the workplace can become quite a talking point. To get them right, you need to make sure that they are more than just a display board with some posters on them. They need to stand out and ideally they should be manned so that people can be engaged in conversation or have a go at whatever it is you are trying to get across.

### Team briefs

These are commonly used on a monthly basis to gather colleagues together to work through a structured conversation. Delivered by the team's manager and with the opportunity for people to participate in the conversation, they are more effective. To help make sure they are, it is advisable to ensure that it is easy for the person leading the conversation, making it clear which bits you want them to say – e.g. *explain* . . . – which bits you want them to use to get a discussion going – e.g. *ask* . . . – and give them some prompts to use in case nobody says anything.

### Organisational magazines

Almost all organisations have an official magazine that they publish for their colleagues. Using it to advance a safety message is one of the commonly used tools we have. However, in truth the effectiveness of them is limited and is linked to something beyond your control – how much liked and relevant the workforce think the magazine is. Like the use of the organisation's intranet, their magazines should be used to back up a message, not deliver it.

### Films

These used to be used few and far between because of the time taken to produce them and the cost. Nowadays, though, this is different as people are more willing to see films that are a little rough around the edges and posted on collaboration sites (see below). The key here is to make them relatively short, probably no more than four minutes and definitely try not to make them your 'typical safety film' with sterile voice-overs and people doing things in a clean environment with clean PPE – they have to reflect real life otherwise people will lose interest in it.

### Face-to-face briefs

By far the best way to communicate is face to face, which is better for the sender and the receiver for many reasons. It can cause problems, though, particularly if you and your team are spread out all over the place. However, there are a great number of tools available now, which means that this is becoming far less of an issue. Here are three examples:

- **Conference calls** These allow large numbers of people to dial into one telephone call at the same time. They are sometimes uncomfortable as you cannot see what the others are doing and you sometimes find yourself wondering if they are listening. Of course, you have to instil some basic call disciplines like not speaking when others are, muting the line when you are not speaking and taking the call in an area with little background noise.
- **Webinars** These combine conference calls with showing slides or similar on the receiver's computer screen. While they have similar challenges to conference calls, they are useful to help get certain messages across, provided you have the right audience.
- **Video calling** These were once reserved for the senior executives of an organisation. However, with the growing availability of video calling technology through iDevices, Android operating systems and others, this is becoming more and more widely used and can overcome the problems associated with conference calls.

### Text messages

Most people have at least one mobile phone and many have two – one personal and one provided by their work. This means that it is possible for an organisation to send text messages to colleagues' phones, enabling a much higher success rate than other methods to help encourage people to take the message in. It is even possible to obtain technology relatively cheaply that sends a message when phones come in certain range of it, meaning that the messages can become even more specific.

### Business collaboration software

Facebook and other such tools are really taking hold in society and with similar collaboration platforms being available for organisations to use as well, this enables organisations to have their own private social networking site. Although there can be some nervousness about using them, many organisations rely on them to help communicate with their people in a way they want to be communicated with and, most importantly, they allow the organisation to receive communications as well.

## Evaluating your success

As with any formula, you need to find out if your answer is correct, and the same applies here – you need to work out if your communications have been effective. Depending on what you could expect to see in the way of movement in some performance measures, realistically in safety such changes can take some time to come through. This means that you need to use other ways to evaluate the success of your communication.

Experience shows that there are three simple ways to do this that tend not to impact on your time:

- Tune in and pick up on the noise or the buzz around the organisation about the subject you are trying to communicate. If you get it right, there will be tell-tale signs that people have picked up the message and it is starting to be remembered. The ironic thing here is that even if what you pick up is negative – for example, you might hear people say something like 'Can you believe they want us to do this now?' – this actually means that your communications have been excellent (as people are talking about it). However, it is what you are communicating that might need more work.
- Ask directly for feedback at safety forums or any other business meeting you go to; you can gain as much from blank looks as you can from animated discussion.
- When you are out and about in the business, ask colleagues what was in the last team brief to see if they have remembered it. Depending on their answer, it also enables you to get into a more in-depth conversation about the subject.

## Conclusion

Communicating a message is a really important part of a safety professional's role. It is far more than just designing a poster and getting it pinned up on every notice board, just as it is more than simply sending an email to everyone and hoping they all pick it up and act on it.

You must think about the message you want to send (bearing in mind there are lots of competing messages) and how you are going to send it. When you consider the pros and cons of each communication tool outlined in this chapter, it is clear that there is no magic bullet. Therefore, it is absolutely key that you use a number of tools to get your message across to your audience in a way they want to receive it and keep repeating it.

# Chapter 5

# Effective safety training

Training is an important subject. It is the way in which employees learn about how they should do the job, developing new skills and knowledge. In health and safety terms training is a critical cog in the machine and is often a key risk control measure. Yet despite the best of intentions safety training is generally dull with lots of mistakes being made in design and delivery for example, there being too much emphasis placed on legislative requirements, having too many slides full of text and going into far too much detail.

At the basic level, training helps us achieve compliance with legislation. However, more enlightened organisations and professionals see it as a chance to improve safety culture and performance. You may think that you will simply outsource the training element; this has advantages as well as potential pitfalls.

This chapter will explore the different learning methods available before considering how to develop and deliver outstanding learning solutions that improve safety as well as showing how you can measure their effectiveness.

## What is training?

Training is one of four learning methods used to encourage people to change their behaviour and cope better with the situations they find themselves in. The other methods are coaching, mentoring and counselling. Training is typically task-focused learning that is delivered in an instructional style – for example, how to operate a forklift truck safely.

### Designing health and safety training programmes

Irrespective of what you intend the training to deliver, there is a general process to follow and this is shown in Figure 5.1. However, experience suggests that most of the time the training intervention is developed in isolation by the safety professional with little thought for the learning styles of those they are aiming the course at, as well as not thinking through how they will measure the effectiveness of the training course. This is important for a host

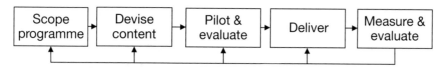

*Figure 5.1* The general process for creating an effective safety training programme

of reasons, not least to justify, in terms other than legal compliance, taking people off their day job for a period of time and being 'non-productive'.

### Scoping your programme

The scoping element of any training programme can be broken down into a number of areas:

*   people's preferred learning style;
*   the training need you are trying to fill;
*   who needs to be involved to help make sure the programme is fit for purpose;
*   how will you measure success.

### Learning styles

In the 1970s two learning and behaviour specialists Peter Honey and Alan Mumford presented a compelling argument that we all have a preferred learning style and, in the main, your preference is one of four styles (1):

*   **Activist**  This means that you generally learn best when you are in a hands-on learning situation – for example, if you are learning how to change the head gasket on an engine you would prefer to learn how to do it by actually doing it rather than reading about it.
*   **Theorist**  This means that you much prefer to learn about things by reading about them, analysing the data and exploring the logic of processes and systems.
*   **Reflector**  These are people who like to take the information they are given, think about what it means and how they can use it, but they do this when they are not in the learning environment – for example, while driving home or walking the dog they often describe it as 'it just came into my head and it clicked'.
*   **Pragmatist**  These are people who love to solve problems by diving straight in and working things out for themselves. For example, they delight in working out for themselves how a stripped car engine goes back together without referring to an instruction manual.

There is absolutely nothing wrong with any of these learning styles. It is important for the safety professional designing a training course to identify what the overarching preference is for the people they are aiming the course at. Many safety courses assume the delivery style that suits a theorist with slides and a few discussion points and exercises, yet this might not be the right approach to maximise the learning opportunity.

There is a learning styles questionnaire based on these four types that you could use to help you work out what each of your colleague's preferred learning style is. However, you can make some reasonably sensible assumptions – for example:

- engineering and accountancy type roles generally appeal to people with a theorist or reflective learning style;
- front-line workers and managers in manual work generally appeal to those with an activitist or pragmatic learning style.

In the modern workplace, safety training should be more than just a course; it should be a training programme which considers more than traditional classroom learning and includes post-'learning' measurement. Hence it is an ongoing process.

There are, of course, a number of other views on learning styles from experience though Honey and Mumford's is easy to understand and apply to people in respect of health and safety and, the argument they presented back in the 1970s still stands up today with many learning and development professionals still using the approach.

### What is the gap?

Unless you have a clear idea of what the training need is, you are never going to be able to work out an effective training course. There are a number of ways to work out what the training need is. Training needs analysis is covered in most health and safety qualifications, but as a reminder you can consider things such as:

- accident, incident and near-miss data;
- results of audits;
- feedback from colleagues at safety forums;
- comparison of training attended versus the defined mandatory training for specific job roles.

Experience shows that it is sensible to keep the scope of the training programme tight, because if you try to cram too much in, it is a hard job to develop it. It is difficult to deliver and crucially people do not remember it all and often go home with a headache!

## Engaging the right stakeholders

To overcome the safety professional writing the training in isolation of the rest of the organisation and other specialists, it is important to get the right people on board to help with the design. In most cases it is good to get at least two other people involved. Figure 5.2 gives an example of who they might be and what each brings to the party. Of course, the level of involvement will be different for different training and the individuals concerned. For a relatively simple subject like risk assessment training for managers, the other stakeholders might just be used to sense check the safety professional's proposals, whereas if you are developing a safety leadership training programme, involvement of the others is likely to be more intense.

## Success criteria

Often we think about determining how successful the training is when we have developed and delivered it, yet it helps to tighten the whole process if from the outset you can start to define what measures you will use to determine if it has been successful and how you will do so. In the main measurement is split into two parts:

- **Underpinning knowledge**   It is vital that people understand the underpinning knowledge of the programme, which can be tested through a written or practical test or a combination of both. Written tests can be marked outside the course, but marking them within the course, with people calling out the answers, does have some clear advantages:

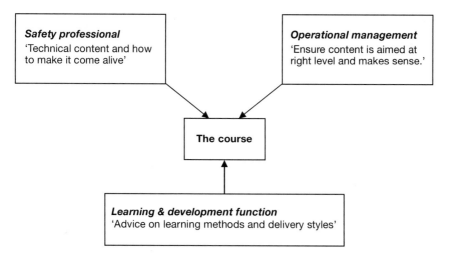

*Figure 5.2* Likely stakeholders in a safety training programme and their contribution to its overall success

*Table 5.1* Example measures of the success of a manual handling training programme

| Measure | Expected change |
|---------|-----------------|
| Employees' knowledge | Employees can demonstrate the right way to lift and handle items |
| Lost time incidents | Reduction |
| Near misses | Increase |
| Stock layout | Changes made reducing the amount of bending and stretching needed |
| Employee engagement | Improvement |
| Peer to peer challenge | Increased level of challenge when colleagues are working unsafely |

mistakes can be corrected instantly, feedback given as to why the answer was wrong, and trends in the answers can be identified which results in the trainer getting instant feedback on their performance and that of the course content.

- **Learning transfer**  Once people have the underpinning knowledge it is important that they take it and apply it in their workplace. This can be measured through monitoring their accident performance, work observation sessions and workplace audits.

Some examples are shown in Table 5.1 that you could use to measure the success of the programme related to manual handling training.

Already from this example list you can see that your expectations are more than just a reduction in manual handling related accidents, so you should go back to your scope and ask whether it is right or whether you are hoping for too much.

## Devise content

The next step is to devise the content of the course or, put another way, think through how you get the technical information across in the most effective way to your audience such that they can remember the content, encouraging them to apply it in their workplace.

Experience says that our default is to put the content in slides using lots of text and then talk through those slides with the odd question thrown in to make the session more interactive. Given what we have already discussed, the success of this approach is obviously limited. To overcome this we need some more tools for us to use, and below are six such possibilities.

### Flip charts

These are great as they give you flexibility in how you deliver; you can either have a series of prepared flips on a subject, you can develop them as you go, you can get everyone to write their opinion or 'answer' on a sticky note and stick them on the flip and talk round them, or even just break things up by getting those on the course to gather round the flip chart for a period.

### Role plays

Experience would suggest that few like role play in the work situation. This said, it is a helpful way to put into practice what you have learnt so far on the course to cement the knowledge you have gained.

### The alternative role play

As an alternative to role play you could simply film a situation and stop the film at appropriate points and ask what is wrong here or what should you do now. As we saw in Chapter 4, this does not have be over-produced or costly.

### Case studies

These are really helpful as they allow you to get people to really think through a real-life problem, how it was overcome and with some well-thought through questions, they can challenge people to think how they would have dealt with it, whether there is a difference and if so why.

### Discussions

Depending on the subject, facilitated discussions are really useful. They work by the facilitator introducing the topic area and then asking some open-ended questions to get the conversation flowing. They then steer the conversation appropriately. From those involved in the discussion they learn through their peers rather than being told what you should do.

To help illustrate this, take the traditional approach to induction training. Typically, this takes the form of 'teach and preach'; the engagement of those attending at best tends to be mixed. Now imagine yourself introducing a few simple questions to the group to discuss fire safety and accident reporting:

- What do you know about fire safety?
- What are the main fire hazards do you think we have here?
- How do you think we manage them?
- If you discover a fire what would you do?
- Why do we need to know about accidents and near misses that happen to our people?
- If you were involved in an accident or had a near miss what would you do?

It might take longer to get the induction delivered, but for something like an induction where the messages will be broadly the same no matter where you go, it enables you to lose the clipboard and cagoule approach of 'I say, you do' to one where you respect people's knowledge and if they get it wrong you can steer them in the right direction.

### Having a go at things

A well-understood training technique is to be shown something once then for you to have a go at it yourself. This works really well for people with an activist learning style. Applying it to practical safety like safe lifting and handling is easy, but as Figure 5.3 illustrates, so too is applying it to more theoretical aspects.

You are the engineering manager at a dairy. On a reactive basis an engineer has to access the top of the milk silos to undertake repairs as well as more regularly planned preventative maintenance of the spray ball and pipe work. The milk silo is 40ft high, it is accessed by a hooped ladder and work is carried out on a slightly domed platform. The main yard is next to the silos and a number of vehicle movements happen nearby.

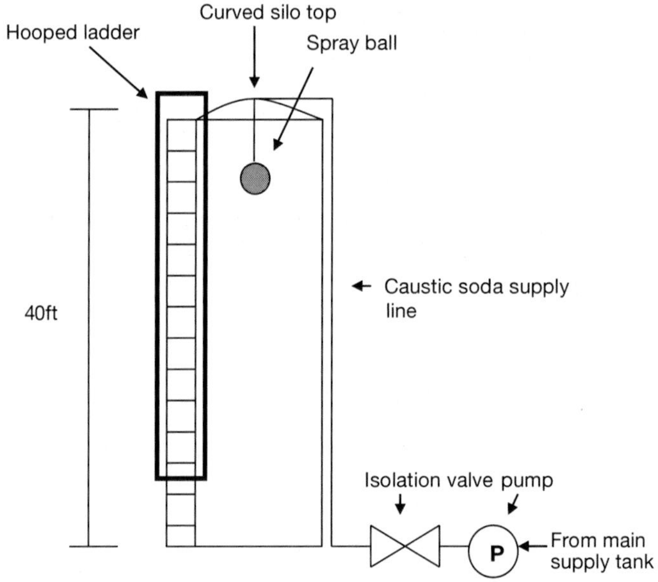

Using this information carry out a risk assessment of the task.

*Figure 5.3* How a case study can be used to help put into practice risk assessment knowledge

### Remember

It is important not to go from using the 'slide and preach' delivery style to just using, say, discussions. The key is to use a range of delivery styles based on the thing you are trying to get across and your audience's preferred learning style.

## Pilot and evaluate

The pilot is about performing a trial of the programme and taking feedback from people about how they felt it helped them to do their job more effectively. As we have already discussed, this needs to seek the opinions of the delegates and must be more than through the normal 'happy sheets' that seem to be given out at the end of most training courses. Being open with the delegates that it is a trial and asking for their feedback is the best option coupled with any feedback they give to their manager post-learning event. Indeed, you should be looking for some early indications that the success criteria you have already outlined are, in fact, starting to be seen.

In an ideal world you would monitor the delegates' behaviour in the workplace to see if they have taken the training on board and acted in the way that the programme set out for them to do. However, in the fast-paced world in which we find ourselves, this is unlikely to happen except with the largest of training programmes. The scoping group should reconvene to review the feedback and agree any adjustments to the programme that are necessary.

## Delivery

This is the easy bit: the product has been designed, tested and proved to work. The challenge with the delivery phase is to ensure consistency in the way in which it is delivered. This is a particular issue where more than one 'trainer' will be delivering the programme. A simple way round this is to devise trainers' notes and for the trainers to sit in on the first few sessions to ensure they understand the way the course is delivered. Dip checks and other monitoring methods are possible and are like those discussed in Chapter 7 in relation to monitoring auditors' scores.

Occasionally, the safety professional will not be the one delivering the programme as it may be left to a team of generalist trainers. In this, the safety professional needs to be satisfied that they are competent to deliver the training. Competency in this case refers not only to how they deliver the training (given their job role you would hope they can do this), but that they have the relevant technical safety knowledge.

Often safety training comes alive with anecdotes and real-life examples that the trainer uses to bring the topic to life. Furthermore, delegates normally end up asking questions slightly off the main content and expect an answer.

While saying 'I don't know but I'll find out' is acceptable, the situation to avoid is the trainer feeling pressured to give an answer and that advice then turns out to be wrong, as this would expose the organisation and the trainer even further. The way round it is for the trainers to undergo some form of safety training themselves if the course content warrants it. Something like a basic site safety induction might not need that level of understanding, although clearly each programme needs to be considered on its own merits.

### Measurement and evaluation

Having delivered the course, as we have already established, it is important to measure how effective the programme has been. For this you should refer back to your measures of success and seek evidence that these have been achieved, depending on the measure this might have to be done in the short, medium and long term – e.g. at +3 months, +9 months and 12 months. Armed with this information you can begin to piece together how effective the training has been, as well as thinking through anything you would do differently in future programmes or, if the current programme is a long-running one, what you can tweak now.

## External training providers

The alternative to developing and running safety training programmes in house is to arrange for an external training provider to do it all for you. Whether organisations decide to go down this route is up to each organisation to decide based on their particular set of circumstances, as one rule does not fit all: some might consider the cost of outsourcing acceptable if it means that their safety professional can get on and do other things, thus 'killing two birds with one stone'. Others might think that there is more value in them getting their own message across.

The following points should be considered when selecting the external training provider:

- Can they deliver accredited courses? Often organisations outsource because they want their employees to undertake accredited safety courses.
- Are they willing to change their standard 'off-the-shelf' product to suit your organisation's specific needs? 'Off-the-shelf' products are just that; they have not been written with your organisation's specific needs in mind. They can often be tailored to major on where you see the organisation's weak points are but you have to ask.
- Determine what you want them to deliver and then contact them. Often you can do an initial filter on the providers you have selected on the phone, from their quotes and how sensible their proposal sounds.

- Once a short list has been determined, meet with them face to face to explore the programme more and work out what makes them better than the next provider. It is useful to take up references from other organisations they have worked with and to see their trainers in action along with looking at their course content. The same rules for an in-house course apply: trainers must be entertaining and knowledgeable and the content must be snappy and relevant.
- When you have determined the provider you want to use, negotiate the price. Typically, the price you are told to start off with is not the cost they will do it for. They will have a price that they can do it for and still make a reasonable margin. If you do not ask you will not achieve what you want.
- Get a contract in place with clearly defined service level agreements. This may sound over the top but from experience it is essential in order to protect both parties. If they are a reputable provider they will already have generic ones drafted and will probably expect you to sign them. Remember to get it checked out by your own organisation's legal team first to make sure that your organisation's interests are protected.

## Conclusion

Training is an ingrained requirement of health and safety. There needs to be a shift away from the traditional view of training being a one-off event. The 'training course' should be part of a much broader programme of learning, which takes the learning from the course and applies it in the workplace. Without doubt experience consistently shows that spending time in scoping a programme properly and considering how you can 'attract' each learning style in your audience pays dividends in the long run.

Outsourcing of training is an option and while it is seen by some as an easier option, there are many pitfalls. Anyone seeking to take this route should follow simple supplier selection and negotiation processes to ensure that they get what they want at the price the organisation is willing to pay.

## Reference

1  www.peterhoney.com/content/LearningStylesQuestionnaire.html (accessed 14 November 2011).

# Chapter 6

# Developing effective health and safety management systems and safe systems of work

To manage health and safety effectively robust systems are required. Since 1999 there has been an international standard that describes a recognised methodology to follow that helps you put in place sound management systems. It is known as the Occupational Health & Safety Assessment Series, often shortened to OHSAS 18001: 1999 (1), and is based around the Plan–Do–Check–Act model. This is likely to be replaced with an updated standard (ISO/BIS 45001) in late 2016 (2).

When designing management systems you really need to be mindful of the end result, the commonest end result is that a management system is put in place. Unfortunately, if this is your aim you have, at least in part, missed the point. As the real end result should be a management system in place *which is followed*. The most common areas in which management systems fall down is that they become over-complicated, impractical and, once designed, the engagement piece is forgotten – in other words, the bit about getting people to follow it. This is not just the case for health and safety management systems (HSMS), though. Experience shows time and time again that the same is exactly true of safe systems of work (SSW), which take the management system down a layer or two and say to workers 'Specifically, this is how we do this task safely'.

This chapter explores how to avoid some of the common pitfalls when designing both HSMS and SSW and, as you will see, the process to come up with them is exactly the same.

## What is a management system?

Very simply, a management system defines the process of control for any given topic, such as product quality, environmental, general business process as well as health and safety. Every organisation will have management systems; some will be good while others won't be; some will be documented while others won't. You do not need to get management systems accredited – for example, the UK's Health and Safety Executive have for many years recommended an approach that, if followed, will help you successfully manage health and safety (3), whereas the OHSAS can be accredited; although arguably the overarching approaches of both are similar. OHSAS 18001 has obvious advantages as:

- It is recognised internationally.
- There is commonality between this standard and others like ISO 9001 and 14001 standards (for quality and environmental management respectively) and so it is very helpful for organisations that have international operations or those that trade internationally as their approach will easily translate; it is worth noting that OHSAS 18001 is nothing to do with the International Organisation of Standards (also known as its short name ISO®) who are developing their own standard (4).
- It is accredited externally and independently – in other words, the complete system (paperwork and implementation) is verified regularly to make sure you are doing what you say you do.

Needless to say, this 'stamp of approval' provides greater confidence in the systems used to manage health and safety but maintaining the accreditation is not always easy particularly if people are not engaged with it. The cost and effort taken to achieve the OHSAS 18001 standard is relatively high, but that said if you have gone to the trouble of designing and implementing a good HSMS, why would you not get it 'badged'? Your tactic should be considered in the context of your organisation: their markets, their appetite, their strategy and their budget.

## What is a safe system of work?

Think of the Haynes Manual (5), which is a great example of a SSW. It outlines the critical (and safe) steps you need to take to affect a repair on a particular vehicle. Not only that but it tells you what equipment you need, shows you pictures of how to do the tricky parts of the task and gives advice on where to go for more information.

Now compare most organisation's SSWs to that description. Experience shows that generally you will be disappointed by what you see. They are typically cumbersome, written in the main to satisfy legal requirements or to help fight civil claims, rather than what they were originally intended for, helping people on the front-line work safely. Of course, SSWs come quite a way down the hierarchy of risk control and rightly so, but they do form an important way in which risk is controlled, so it is important to get them right and, like establishing a HSMS, it is crucial to get them written down and followed.

## When do management systems and safe systems of work really help?

The measure of a good HSMS or SSW is that they are followed, yet it is unlikely to be the experienced worker or manager that refers to them, as they tend to know what to do. They really come into their own:

- where the situation is novel;
- for a new starter;
- when something has gone wrong and you need to work out why;
- when developing training (see Chapter 6).

The paradox, then, is that if you get worker training right (in part based on a really sound SSW) and have robust HSMSs in place you are less likely to have to defend your level of legal compliance or fight civil claims as your organisation's safety culture and performance will be improved. Yet while people can see this link, having the faith to break that vicious circle is challenging.

## Getting them right

As discussed previously, while the outcome might be different the actual process to getting really effective HSMS and SSWs is the same and can help to break the vicious circle mentioned above; it can be broken down into three distinct parts:

- get the right people involved;
- identify the right things to focus on;
- explain things in the right way.

### Getting the right people involved

Despite common practice, HSMS should not be written by the safety professional in isolation then directly issued (or passed on to someone else to read and comment on before being issued). This way of working does not encourage ownership. In a similar way, taking a template from the Internet or other sources and inserting the organisation's name is not necessarily the right thing to do either. There are clear reliability issues with the information they contain, especially if you do not know if the source is trustworthy.

A criticism often levelled by front-line workers about SSWs is that they are unworkable in the real world, often for similar reasons as HSMS are not as good as they could be. The way to overcome this and, to drive engagement from the outset, is to establish a mixed working group of the right people. The right people are those who either do the job or have an interest in the topic area; some examples of whom they might be are

*Table 6.1* A sample of HSMS topic areas and likely interested parties

| HSMS Topic | Interested parties |
| --- | --- |
| Hand arm vibration | Procurement<br>Occupational health<br>Human resources |
| Fire safety | Property/facilities<br>Training<br>Fire alarm and emergency lighting service provider<br>Fire extinguisher service provider |
| Asbestos | Property/facilities<br>Asbestos survey service provider |

*Table 6.2* A sample of SSW topics for a typical manufacturing plant and likely interested parties

| SSW topic | Interested parties |
| --- | --- |
| Cleaning food production areas | Operatives undertaking the cleaning task<br>Quality assurance team |
| Removing blockage from decrating machine | Operatives who have to remove the blockage<br>Engineering team |
| Replacing a drive belt on the in-feed conveyor | Engineering team<br>Engineering manager |

shown in Tables 6.1 and 6.2. Hopefully, nowadays it goes without saying that employee representatives should be included in the discussions as well.

The role of those involved in this process is to actively contribute to what the HSMS and SSW will look like, the safety professional's role is simply to steer them in the right direction and challenge their thinking about anything that might not be safe, or not necessarily the right thing to do. But experience shows that this is likely to be the exception rather than the rule. After all, 90 per cent of what people want to do will probably be acceptable from a health and safety point of view; it is only the remaining 10 per cent when the safety professional may need to be more direct.

### Focusing on the right things

This is where you start to work out the actual content of the HSMS and SSW. Out of all the areas you need to cover, applying sound risk-management principles helps you prioritise the things you should be looking at first.

As a guide, those that pose the most risk as well as those that have been involved in more incidents should feature towards the top.

Now you have the right people in the room and you know the areas you are looking at, you can start to get into the detail of things. Having explained what you are there to do, asking the following questions will start to help you get the necessary information you need:

- What is it we are trying to do?
- What are the key steps involved in doing this?
- Which are the steps that we often get wrong or don't follow?
- What can we change so these steps are right and followed?
- How do we check to make sure we are doing what we say we will?

Now let's look at some real situations to help bring this approach to life. In Table 6.3 the HSMS is looking at putting in place a defect reporting system and the SSW is how to change a wheel on a car safely.

The other good thing about this exercise is that you also have a list of actions for the organisation to do to help embed the outputs and improve their overall assurance – e.g. in the case of the example in Table 6.3, auditing the asset register.

Having completed this first exercise with the group you can start to work through more of the detail, with the example of developing a SSW for changing a car wheel you could ask them to expand on each of the critical steps. For example under the *'Preparation'* heading you might get the following steps:

- Make sure the vehicle is on flat level ground.
- Double chock one of the wheels that will remain on the ground.
- Visually inspect the jack ensuring there are no obvious defects as well as the axle stand.
- Place the jack under the car's jacking point.

Under the *Raise the vehicle and support* heading you might get:

- Operate the jack to lift the car.
- When the jack starts to lift the car, make sure it is still under the jacking point. Do not lift the car too high, only raise it enough so that the tyre is off the ground.
- *Always* place an axle stand under the car's chassis. If this is not possible, place it under the car's sill but do not lower the jack – the axle stand is only there in case the jack fails.
- *Never* go underneath a car that is only supported by a jack.

Keep doing this until you have covered all the key stages and you will have the outline of your HSMS or SSW. Then, having done this, revisit them

Table 6.3 How this approach can be used to develop an HSMS for defect reporting and a SSW for changing a wheel on car

| Question | HSMS: defect reporting | SSW: changing a wheel on car |
| --- | --- | --- |
| What is it we are trying to do? | A system to report a defect on any plant or equipment used across the business. | Changing a wheel on a car in a garage work shop. |
| What are the key steps involved in doing this? | Define what a defect is. Who/how do we report a defect? Defect log. Process for resolving the defect. In-house engineering. Arrange external contractor. Process for scrapping the item. Update asset register. Order new item. Training and communication. | Prepare the area and get kit. Raise the vehicle and support. Replace the wheel. Lower the vehicle. Final checks. |
| Which are the steps that we often get wrong or don't follow? | Updating the asset register when we scrap items. | We don't always use a secondary support, we rely on the jack. |
| What can we change so these steps are right and followed? | Improve training/ communication to responsible managers. Audit the asset register. | More monitoring by managers. Make the message clearer in the SSW. Always use a secondary support. Show a picture of the right way. |
| How do we check to make sure we are doing what we say we will? | Include a check in monthly manager self-assessments. Make feature of annual audit. | Use of near-miss reporting. Monitoring by managers. |

and decide which bits do not really add anything and strike them off or, if you think they all add something, try to condense the list. Remember to refer back to the relevant risk assessment or piece of legislation to make sure you cover all the bases.

## Explaining things

What we have discussed so far is not earth shattering – there are only so many ways you can develop an HSMS or SSW. What will set it apart and, where really big wins are for worker safety, is how the information is presented. Explaining things in the right way will encourage more people to refer to them. This is so important that Chapter 4 is dedicated to this subject but here are four specific ways to effectively communicate HSMS and SSWs.

### Photos

Adding a photo into a SSW helps to draw people's attention to the thing the picture is showing. Take, for example, the comment in Table 6.3 that sometimes the secondary support (axle stand) is not always used when the vehicle is in the raised position. A picture showing the correct use of the jack and axle stand in the SSW helps to tackle this problem as it shows what good procedure looks like. This means that everyone knows what the safe way of working looks like which removes the 'I didn't know' argument from workers who don't follow the rule and, managers have something to compare things to when they are monitoring whether people are doing what they should be.

### DIY-style diagrams

Putting flat-packed furniture together used to be challenging 15 years ago until the quality of the instructions improved no end. Nowadays, the instructions are generally pictorial and really easy to follow as they show you each stage, the tools you need and zoom in on key areas. Safety professionals can learn a lot from this simple yet highly effective approach. Imagine giving front-line workers a pictorial guide on how to do the task rather than having a SSW with lots of words, they could understand it irrespective of whether they can read or not. This is particularly helpful where you have migrant labour or people with a low reading age.

### Flow charts

Flow charts can be used very effectively for either HSMS or SSWs as they show the process flow for doing a task and where there is a choice to be made show the different options. Much like using DIY-style instructions, these help remove the need for lots of words and enable people to work through the different options given the circumstances they find themselves in.

### Films

As we have seen in Chapter 4, technology today means there is no reason at all why a short film cannot be made with somebody actually doing a task following the SSW or HSMS, explaining what they are doing as they go along. Experience shows that when this approach is taken, engagement, knowledge retention and 'compliance' increases. To see the style of these films for yourself, go on to an Internet search engine and type in 'how to lay a patio', then on the results page click on the video tab – there will be lots of short films for you to watch.

### Location

Having developed your HSMS or SSW you need to think about where best to locate it to encourage their use; it is about making sure they are available when they are needed. However, there is a compliance element that needs to be considered to achieve this part; they are often put in a manual or on the organisation's intranet.

However, to get them to help you improve your organisation's safety culture and performance having the more frequently used ones stuck up in a prominent position is helpful provided they do not become 'wall paper'. Take a management system for accident reporting; an example of where sticking it up in a prominent position works well is if the system has changed or if there is a high number of people who might need to follow it, but they tend not to deal with it on a regular basis – e.g. a weekend supervisor in a shop.

Similarly, the best SSWs are those that are a bit dog-eared or dirty, as it shows that people use them. Think about a SSW for the safe use of a machine. It is best placed on the machine by the controls where the operator stands and the one that refers to dealing with a blockage is best located where the block occurs.

### Engagement

Often engagement is seen as a separate piece of the jigsaw – i.e. you develop the HSMS or the SSW and then seek to engage people with it. However, the approach outlined in this chapter helps you develop the HSMS or SSW while engaging people with it from the beginning, meaning that the process is slicker and helps you get something that is workable and more likely to be followed. Not only that, but if someone has been involved in developing it, they are more likely to encourage their colleagues to follow it, as 'it is theirs'.

### Conclusion

Clearly defined effective HSMS and SSWs are vital components in any organisation's approach to improving their safety culture and performance. Designing them traditionally has been viewed as the safety professional's 'bread and butter' and something done in isolation with little thought of the end result other than a tick in the box.

To help ensure that they are implemented and followed, it is important for the safety professional to get a real-world perspective and to do this, it should be opened up to a wider organisational audience for their input throughout. While this will help improve engagement and ownership of the process, it can also help identify possible efficiencies in the way the organisation manages aspects of the business.

Developing these, though, is only part of the job. Getting people to follow it is vital and this is probably the hardest thing to do; without it, you will just measure how unsuccessful they are. Both HSMS and SSWs should be simple to use, easy to follow and their contents acceptable to all – health and safety is bad health and safety if it gets in the way of doing the job.

## References

1   www.bsigroup.com/en-GB/ohsas-18001-occupational-health-and-safety/ (accessed 14 November 2015).
2   www.bsigroup.com/en-GB/ohsas-18001-occupational-health-and-safety/ISO-45001/
3   www.hse.gov.uk/pubns/books/hsg65.htm (accessed 14 November 2015).
4   www.iso.org/iso/home/standards/management-standards/iso45001.htm (accessed 14 November 2015).
5   https://haynes.co.uk/ (accessed 14 November 2015).

# Chapter 7

# A model for auditing

Auditing is an important part of any management system, but maybe more importantly audits provide the opportunity to give confidence that the business is doing what it says it will do. In essence, audits are just a tool to help provide assurance (freedom from doubt that you are managing your risks effectively) and that their roots lie deep in corporate governance. There seems to be a growing industry providing health and safety auditing services and although undoubtedly effective, most companies would probably like their own bespoke auditing programme, given half a chance.

A common dread among operational colleagues, it seems, is that of 'being audited' for a whole host of reasons:

- occasionally their results are used 'as another stick to beat them with';
- they are perceived as taking up lots of time;
- to some they are thought to be more hassle than they are worth.

Irrespective of these views, audits are here to stay and opportunities exist to use them not only to provide the necessary assurance and highlight areas for improvement that good corporate governance dictates, but also to increase stakeholder engagement, safety performance and develop the organisation's safety culture at pace. This chapter outlines how to design an audit programme and suggests an audit model that organisations can 'pick up and use' quickly.

## What is an audit and its key components?

In its simplest form an 'audit' is a technique used to test how robustly managed an issue is compared to a given standard. That standard could be legal requirements, organisational policy or industry best practice. In other words, they measure your performance against defined criteria and provide feedback on how well you are doing as well as setting future direction.

## The basics

In the main, the majority of audit programmes will have three basic components:

- an audit protocol;
- various sampling techniques;
- high validity and reliability.

## Protocols

Audits are undertaken to a set protocol. These outline what and how you audit and are typically a question set. There are many 'off-the-shelf' audit programmes available and when used to their potential are extremely impressive. However, in deciding whether to use such programmes you should be mindful of their advantages and disadvantages (examples are shown in Table 7.1) generally associated with this type of solution.

Experience shows that, given the choice, the majority of organisations would prefer bespoke audit programmes. Unfortunately, they are often put off by the perception that they take a lot of time and effort to write, or they do not have the ability 'in house' to do a good job. Realistically, they are not that difficult; all that is needed is an understanding of the topic being considered and a logical mind.

Table 7.1 Some of the advantages and disadvantages of typical 'off-the-shelf' audit programmes

| Advantages | Disadvantages |
| --- | --- |
| The question set, guidance to help determine level of compliance and the actions required for a deficiency have already been prepared. | Most programmes require an initial fee and then a fee is payable per licence (for each user). Updates and support can be costly. |
| Software audit programmes tend to be linked to a database, enabling very quick analysis of the data. | Often the software packages can do far more than the organisation actually needs and thus significant sections are not used reducing the return on the investment. |
| The provider will often offer ongoing technical and user support as well as continual development of the programme to take into account new developments. | Limited ability to make the programme organisation specific presents validity problems. |
| Improved customer confidence when auditing to an independent standard (depending on how well known the 'off-the-shelf' solution is). | Tend to be generic and do not lend themselves to more unique industries. |

Let's take the relatively simple example of the Health & Safety Policy Statement. In the UK the standard to which you want to determine your compliance on this subject is the Health & Safety at Work etc. Act 1974, specifically Section 2, Part 3. This states:

> it shall be the duty of every employer to prepare and as often as may be appropriate, revise a written statement of his general policy with respect to the health and safety at work of his employees and the organisation and arrangements, for the time being in force for carrying out that policy, and to bring that statement and any revision of it to the notice of all his employees. (1)

The question set to determine the level of compliance with this requirement would generally include the following:

- Is there a Health & Safety Policy Statement signed by the managing director (or similar) within the last twelve months?
- Does the Policy Statement outline the company's general position in respect of health and safety?
- Is it displayed in a prominent position at company locations?
- Have all staff read it and are they aware of its content?
- Is this formally recorded – for example, on their Employee Training Log?

Obviously, more complex issues may take more time to think about, but the basic concept remains the same.

### Sampling techniques

A good audit is more than just taking things at face value. It is all about getting formal evidence to qualify the answer. Different sampling techniques are used to gain the data upon which you make your decision about how the area being audited is performing. What is important is to test things from different angles and Figure 7.1 shows how the idea of triangulation discussed in Chapter 3 can be applied here.

Typically, there are three sampling methods used to help build a picture of how things are managed:

- **Interviews** with managers and employees are commonly used and are helpful in getting the necessary information (anecdotal or otherwise). However it is worth remembering to go into them prepared; their time is precious and also think about whom you are talking to – for example, your approach should differ when talking to the site manager which might be 'pre-booked' and more formal while an interview with an employee would likely be more of a chat and ad hoc.

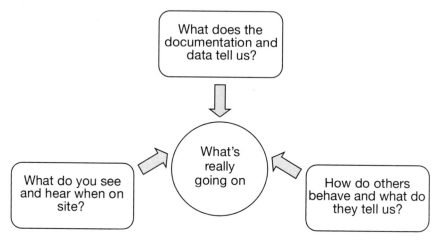

*Figure 7.1* Triangulation model of data gathering and validation

- **Inspections** are used to perform physical checks and observations of working practices. From experience, the trap to avoid here is to work off a 'tick list'. Doing so can make the person undertaking the audit become blinkered and not pick up on the real problem because it is not on the list. Health and safety these days should not be about tick sheets. That said, it is understandable that people want to have some structure to their work and the production of an aide-memoire to point people in the right direction, helps to right the balance of the tick box approach.
- **Reviewing data and documentation** helps show whether the approach you have been told about and seen in action is formally recorded. Looking at typical safety performance measures like accident records and enforcing authority activity also helps determine if things are being managed as they should.

### Validity and reliability

It does not matter how much work is put into the audit protocol and the collection of the data, as the audit will lose all credibility with operational management and fail to provide adequate assurance, if it does not appear to test what it claims or if the results are 'questionable'.

A major factor in achieving good face validity is that it is vital that the protocol is developed correctly. Once developed, it should be tested at a number of locations to see if the audit delivers the expected results. To overcome any bias you need two people to do this – one to use the protocol, the other to determine compliance with the standards in another way.

Reliability is about ensuring that the same results would be achieved if the exercise were repeated. This tends to improve when there are smaller numbers of auditors involved and defining it can be challenging as audits are just snap shots in time. What you need to avoid, though, is extreme variance in people's marking. While having a well-defined audit protocol will control some of this, you also need to control the auditor's individual differences in interpretation – for example, some people might be extremely harsh in their view or too lenient. This can be done by undertaking 'levelling', of which there are two main types:

- **Physical levelling** is where all the auditors are asked to audit the same location at the same time, following which there is discussion about the results leading to a standardised interpretation. This interpretation is then applied to all future audits.
- **Desk-top levelling** works where numerical scores are assigned in response to the level of compliance with each audit question or category. The exercise involves determining the average score for each auditor against each of the questions or categories. If all the auditors are scoring within +/–5 per cent of the mean average score of all the auditors, then that is generally accepted as the auditors being consistent with their interpretation of the question and the situation presented to them. Any significant deviation from this would suggest one of two things: either the auditor(s) is not being consistent or the locations are significantly different in the way they apply the application of the organisation's health and safety standards.

There are clear pros and cons for each methodology, the physical levelling is really useful when starting the audit process to help get the auditors on to the 'same page' as each other, while the desk-top option is less labour-intensive to start off with, but it may require pulling the group together to carry out a physical levelling session as a result to understand differences and agree a standard. Best practice suggests that it would be sensible to start off with the physical levelling and then review the auditors at a predetermined interval with the desk-top review followed by a physical levelling as necessary, but probably at least every six months.

## A robust audit programme

We have already established that auditing is a given and on this basis resources will be committed to it; done correctly, there should be a positive impact on the organisation's safety performance and culture. To help make sure you really do get as much out of the audit process as possible, a best practice audit programme, used by many organisations, would be made up of six separate but linked elements. The programme has a number of merits:

- it frees up significant amounts of time for the safety professional to provide support where it is needed most (the poor performing sites) and other safety improvement programmes;
- it greatly advances the safety culture at a quicker pace than traditional audits;
- it still gets the job of auditing done.

### Organisational management and control

This considers the effectiveness of the organisation's policies and management systems at achieving compliance with a defined standard and so helps to identify the cause of problems which, if left unresolved, present themselves as accidents or incidents on the front line. Depending on the size of the organisation, these can be undertaken by a safety professional or an in-house general auditor who might not know much about the detail of safety but understands how to audit. There is no issue with this provided the protocol they are working to is well defined by a competent person beforehand. How often this sort of audit is undertaken depends hugely on the needs of the organisation and the levels of change made to the policies, procedures, legislation and industry guidance.

### Operational High Risk Activity Control

An Operational High Risk Activity Control audit collects data about individual site management of 'high risk' obvious health and safety issues allowing such things to be detected and corrected quickly. This approach deals with the symptom of the problem rather than the cause (which is picked up by the Organisational Management and Control audit) and helps line managers control a situation to prevent an accident or incident. It is based on the assumption that most organisations have a 'compliance'-type auditing function which visits multiple locations to measure their compliance with organisational procedures on various issues linked to front-end administration and financial management. There are clear synergies between this group of people and the work of the safety auditor; such synergies can be exploited to create capacity for the safety professional to do other things. This type of audit simply builds on the existing Compliance Team Audit with the addition of a number of health and safety related questions and uses other people to be your 'eyes and ears'. Take, for example, a food retail shop where one of the highest risks in the operation is associated to customers being exposed to the hazard of slips, trips and falls on the same level. The protocol

---

**Statement:** Slips, trips and falls on the same level are adequately controlled.
Scoring: Yes = 10, No = 0.

Marking Guidance: Answer **'Yes'** if all of the following, where relevant, appear to be in place:
- no spillages or other debris on the shop floor;
- anti slip mats are located at the entrance to the shop and at the fruit and vegetable aisle;
- employees know what to do in the event of a spillage;
- wet floor signage is available;
- no pot holes or similar on the shop floor;
- variations in floor levels are highlighted;
- display plinths are not left without anything on them;
- stock is not stored on the floor.

*Figure 7.2* Extract of an Operational High Risk Activity Control audit protocol for a food retail premise

for the auditors to follow would be along the lines of that noted in Figure 7.2; equally, you can build up a score that is less clear-cut than shown such that for each control the location has in place they get a point.

The critical part of this audit is that you are keeping the obvious high-risk safety issues at the top of people's agenda and for this reason it is ideal to keep the protocol to around 10–15 key areas. For each possible deficiency noted, a standard set of the corrections is developed, making the life of the auditor easier. It is also important that there is the opportunity for additional hazards not covered in the audit to be highlighted along with the necessary corrective action, as it would be remiss of the organisation to become blinkered and to consider only what is on the protocol.

Naturally, the people carrying out the audit will need some training on what to look for and most importantly they will need to know what to do should they be unsure about something: stop and seek advice from a safety professional. The only real drawback to this approach that you need to be mindful of is that if they are undertaken by a non-safety professional they tend to be very clear-cut in their determination on how something is being managed rather than taking the more pragmatic risk-based approach that hopefully safety professionals take given their understand of legislation, associated guidance and knowledge of risk-management principles.

### Coaching and development

If the Operational High Risk Control Audit creates capacity for the safety professional while maintaining impetus, the 'Coaching and Development' audit allows you to focus resources where they are needed the most – the poor performing locations. However, how do you know where your worst performing sites are? While you may have a 'gut feeling' about a handful of them you might not know them all that well, particularly if you have a large number to look after – hence the need to develop a risk profiling tool (discussed in Chapter 14) that uses a blend of objective and subjective data to create a list of high-, medium- and low-risk sites.

Once you have determined the organisation's risk profile, the hard work begins: supporting sites to improve their performance. Simply put, the Coaching and Development Audit's purpose is for the safety professional to coach and support the site manager to enable them to understand what the problem is and how to implement systems or other control measures that will work at that site to ensure that their safety arrangements are adequate and performance improved.

Coaching for safety performance is such an important tool for the safety professional that Chapter 15 focuses entirely on how to do it effectively, but the general approach in these circumstances should be: you arrive on site and walk round with the manager helping them see that the way they are managing things now could be improved, then as you go you start to ask them how they could improve them. By the time you leave the site you will have helped the manager create an action plan (that you can formalise for them) by facilitating them, saying what the problems are and what can be done about them. This sort of approach has a number of advantages:

- It enables a firmer direction to be given by the safety professional to any really high-risk area that is not adequately addressed, hence reducing the businesses exposure.
- It helps make sure the risk controls put in place are workable (as the manager with their team have come up with the outline for them).
- It provides hands on real-life development for the manager that does not require them taking a day or more out of the business for training.
- It changes the clipboard and cagoule perception of safety professionals.

The key to success with this is not to think that the work is finished when you leave the site. Integral to coaching and development is feedback. Thus, the plan to return to the site at a mutually convenient time, say, in six weeks, to see how the manager is getting on. This enables you to ensure that they are going along the right lines, offer further coaching, feedback (both developmental and offer positive praise) and, in the worst case, provide evidence to their line manager that they are not doing what they should be, thus making safety a performance management issue (see Chapter 20).

### Self-assessment

As the safety culture of the organisation develops, or you feel it is time for operational managers to engage more with health and safety, a self-assessment can be used. Like developing any audit, consider what areas you want the managers to look at and then develop the question set accordingly; in this case it could be linked to the more problem common areas noted in the Coaching and Development or High Risk Operational Control Audits. The self-assessment should not just be looking at paperwork; it should be a mix of both observations of the working practices as well as a paper-based management review. There are four key things to remember though with using self-assessments:

- They can become a joke among managers, insomuch as it is a tick-box exercise, so they should only be used when the safety culture has developed enough, so that the chances of this happening are reduced or organisational systems are in place to check the validity of the audits.
- Managers have many other things to do and priorities to consider as well as managing safety, therefore whatever you ask them to do should be broken down into penny packets or manageable chunks. Not only does this approach make it more likely that they will do it but it will also be easier to administer.
- How often you ask managers to complete a self-assessment will determine how effective they are at providing assurance and improving safety. Some organisations ask their managers to check two things a week so that over the course of the year 104 safety checks have been completed. You might think that is too much and quite hard to implement, but once a quarter might not be sufficient and monthly may be more beneficial for the organisation.
- Do not forget that as this takes off you will get a lot of requests for your time to help put things right, so another important factor in determining the frequency is the level of resources available to help solve problems highlighted and share best practice; thus, it might be more sensible to consider the 'big ticket' items.

While this is aimed principally at engaging front-line managers, it can also be used to revitalise a Health and Safety Committee Meeting, utilise more effectively safety representatives and develop team leaders.

### Senior management review

This is concerned with getting senior managers and directors to have a structured and systematic conversation with front-line employees about site health and safety practices when they are out and about. The protocol should be kept to four or five key questions that are clearly linked to the organisation's overall health and safety strategy as this will help to show senior management commitment. The questions should not be detailed, complex or technical. Organisations that use this approach simply aim to demonstrate to front-line staff and managers that senior management take health and safety as seriously as other business concerns.

To help get over the idea of the clipboard and cagoule mentality towards health and safety, it is important to include one question to encourage the identification of best practice – for example, 'What is the one thing you have seen here that really helps improve safety that you haven't seen anywhere else?' At the end of the exercise the senior manager or director should talk the manager through what they have found and the manager should 'contract' with them to put right the necessary points. This on its own might be seen as too weak a response, so the findings are often shared with the line manager's immediate boss for them to make sure the manager follows through on the actions accordingly.

### Grab sample assessment

This can be undertaken by anybody, following a small amount of training to show them what they are looking for and can be used to rapidly build up a picture of compliance with a specific issue at a large number of sites at the same time. For example, as part of the organisation's occupational road risk programme you might want to check that all vehicle drivers leaving your site are wearing their seat belts. The people doing the checking could be from any part of the organisation – e.g. Finance, Human Resources, Marketing have been briefed about what they are to note (in this case how many drivers are wearing seat belts while in a moving vehicle) they can undertake the review at a set time. A set time is critical, otherwise the first site visited might tell their neighbouring ones to be 'on their guard' and thus the picture you get will not be reflective of reality. The idea of this audit is to take a grab sample across a number of sites at the same time on the same issue. It is also supportive of developing a safety culture into non-operational functions, as it helps develop such people for their own personal development or who might have to go to sites from time to time to understand and appreciate the safety rules they should follow when they are there.

## Conclusion

Audits form a major and, all too often, time-consuming part of the safety professional's job. The key to success is to make their emphasis a safety improvement tool that provides assurance to the relevant stakeholders, rather than a compliance check. To maximise the opportunity audits present, organisations would do well to consider doing more than just a one-off annual audit and take a more holistic approach, which draws more people into the process to help spread the load and improve safety culture at the same time.

Although links have been made between elements of the auditing model, those not explicitly noted should be apparent. Each element is linked and all feed into the bigger picture which provides the necessary assurance (or otherwise) to the organisation that it is doing what it says it will do.

## Reference

1    www.legislation.gov.uk/ukpga/1974/37 (accessed 14 November 2015).

# Chapter 8

# Developing a business case

It does not matter where you work, or what your organisation does, cash is king. In the private sector if you don't make a profit in the long run, you don't have a business, and in the public sector if you don't have enough money in the pot, you can't provide the services you are required to. This means that while there are moral and legal considerations that drive organisations to manage and improve their safety culture and performance, the biggest driver tends to be financial. This has to be more than saying 'If we do this we will reduce accidents and so save on sick pay and insurance claims'. You need to be able to translate this into a tangible cost and benefit to create a sound business case. The ability to do this is a fundamental skill that all safety professionals need, yet it is surprising how many cannot do this effectively. In part, this is because safety qualifications tend not to cover how to do this and it can be difficult translating safety benefits into monetary terms unlike three areas that are linked to the safety professional's role:

- **Driver safety** Visibility of the costs associated with vehicle damage are readily available from an organisation's insurers and require little manipulation to make them easy to use in a business case to invest in better driver risk management.
- **Occupational health** Sickness absence and the associated costs are generally well monitored by an organisation's HR functions. They use these to help show how much value they add to an organisation by reducing cost through getting people back to work quicker. It is also easier for them to show how investing in an occupational health case management system with an external provider can reduce the sickness cost even after paying for the service.
- **Environment** The costs related to waste disposal in many organisations are relatively high and well known. This means, like the driver safety example above, that it is easy to translate this in a business case to recycle the waste such that it becomes cost neutral or in the most effective cases a revenue stream.

In this chapter we will explore how you can create a sound business case for safety, showing how to translate safety data into real cost and benefit, as well as how to effectively present the more intangible benefits associated with safety improvements.

## The basic principles

It is always an easier conversation to seek investment in safety when you can demonstrate a clear positive gap between the cost of doing something and the benefits you expect to achieve. To find out if this is possible, there are a few ways you can do it. However, in this chapter we will look at the most commonly used and easiest to do method. For this you need to work out a) the cost of action and b) the cost saving or benefit.

## Working out your costs

This really is the easy bit; you need to work out how much your safety improvement programme is going to cost to do. You can get this from either obtaining a quote from your external service provider or building the programme costs yourself, just as the way you work out your monthly household spending.

## Working out the cost saving or benefit

This element is a little harder to do, but only because you have to translate safety in monetary terms; to do so, you need to work out how much a lost time accident costs your organisation. To do this you need to know two things:

- The average number of days taken off per accident in the last twelve months.
- The average salary of your front-line colleagues (or whoever has the accidents in your organisation).

By multiplying these two figures together you get the average *direct* cost of lost time accidents in the last twelve months. However, this does not take into account the other *indirect* costs associated with accidents like management time spent investigating it, the cost to make good or alter plant, equipment or buildings as a result, the increase in insurance premium or the costs associated with minor accidents. Some years ago the UK's Health and Safety Executive developed a cost of accidents model (1) based on research which found that for every £1 of direct costs you could attribute to an accident there were between £8 and £36 worth of indirect costs.

With this in mind, you need to find out if your organisation operates a really tight cost base, in which case you might want to multiple the average direct cost by £6 to give you the true cost or, if it is a more expensive set-up, you could use a multiplication factor of £36. Unless you are really sure what your organisation's cost base is – as the worst thing you can do is 'over-egg the pudding' as you will lose credibility before you start – go for just below the midpoint and use a factor of £20, which will give you a conservative figure. Now you simply multiply the number of lost workdays your organisation has had to do the specific topic you are looking at to show how much they have cost.

## Worth doing?

If the amount you expect to save (the benefits) outweigh the cost of doing the safety improvement action, there is a good case to progress. Here is an example to illustrate how this approach can be used.

Imagine you are the safety professional responsible for a site that has been having a large number of lost time accidents relating to people slipping on the terrazzo flooring. Working with the facilities team, you have identified that the problem is that the floor needs a regular deep-clean regime that will cost £25k per year. You work out that there have been four lost time accidents over the recent 12-month period that have resulted in 80 days being taken off work as a result. The average salary of those involved is approximately £19k per year. Table 8.1 shows the steps of the calculation you take to determine the amount the accidents have cost.

As the table above shows, the benefits of the deep clean potentially will improve safety and deliver a significant cost saving.

### Half and half

If the gap between your costs and benefits is not that good, do not give up just yet – you just need to do more work to build the business case. Look to find more benefit elsewhere – for example, is there a cost of damage associated with the accidents or process connected to the accident that, if added into the equation, would help make the case? While it is preferable

Table 8.1 The cost of accidents caused by the poor floor surface in the example outlined

| Costs | Calculation | Amount |
|---|---|---|
| Average daily employee salary | £19,000/365 | £52.05 |
| Cost of lost work days | 80 x £52.05 | £4,164 |
| Apply hidden cost multiplier | £4,164 x 20 | £83,280 |

to have monetary benefits alone at this stage it is acceptable to have a blend of tangible and intangible benefits.

Let's use another example to explore this further, a distribution firm has operatives picking a number of goods that are then put on a pallet and manually shrink-wrapped. As part of the safety professional's safety improvement programme, they have identified that if the shrink-wrapping process were automated it would reduce the wear and tear on colleagues' bodies and therefore reduce the cumulative effect of manual handling. They understand the cost of buying the machine and the ongoing costs, but they have not had any lost time accidents directly associated with the process.

Of the eight lost time accidents the organisation has had in the last twelve months associated with manual handling, they estimate that reducing the wear-and-tear element on the body of manually shrink-wrapping could save year on year at least one such accident. Unfortunately, having worked out the expected monetary benefits their case was slightly out – in other words, the cost outweighs the benefits. They know they need to find some intangible benefits to help tip the balance in favour of the improvement. The idea for the machine came from some front-line staff, so trialling their suggestion would help the organisation with their employee engagement journey and, following a call to the supplier, they can have the machine on a three-month trial.

Suddenly the business case is more solid. The key to extending the trial is to do a really good evaluation; in this case you could ask the following questions. Have you:

- reduced pressure put on colleagues' bodies by less bending, twisting and stooping;
- improved operational efficiency, if so by how much;
- reduced the cost of shrink-wrap brought;
- improved load security;
- achieved a more professional-looking pallet?

## The intangible business case

Arguably, the inference until now has been that a business case built on intangible benefits is to be frowned on. However, there are times when this is the only way forward. Where this is the case, it is really important to use a wide range of realistic measures that people can look at and can see that they are valid and not you clutching at straws.

Pretend for a moment that you want to run a two-day programme that will take two colleagues per location out of their day job so that they can go back and change their workplace layout to reduce the amount of lifting and handling people have to do. Ahead of seeing the operations director, based on a series of conversations with their team and sense checkers (see Chapter 16), you note down the expected benefits you think this programme will deliver. They include:

- reduced manual handling related incidents;
- improved customer service as colleagues will be able to find products a lot easier;
- increased operational efficiency because if things are organised in such a way as to reduce manual handling, it will mean that colleagues will not be running around all over the place;
- increased employee engagement as people will be coming up with their own solutions to problems rather than being told what to do.

Hopefully, reading these measures now and comparing them back to the other approaches discussed, you can see how weak they seem. Organisations are more accepting of this approach, though, when you want to take people out of the business for a day or two, but not if you are going after relatively large sums of money.

## Conclusion

Experience suggests that safety professionals lose credibility when they seek money for improvement programmes; in fact, quite often you hear them berating their organisation for not giving them the budget they wanted. The truth is that everybody has to fight for resources, even in times of economic boom and arguably if you cannot translate safety improvements into monetary terms you will lose out.

## Reference

1    HSE, *The Cost of Accidents. HS(G)96* (1996) Sudbury (out of print).

# Chapter 9

# Report writing and presentation skills

It does not matter how good your idea or investigation, you need to be able to continue that credibility when you are communicating your ideas. Partly due to the career paths that safety professionals follow, many struggle with this. In this chapter we will look at how to complete strong reports and powerful presentations, both of which are more than just writing the words or preparing the slides with animations.

## Starting right

Both a report and presentation are systematic and structured ways to explain a problem, situation or an idea. This means that irrespective of whether you are writing a report or preparing a presentation, the basic approach is the same and links back to the concepts we discussed in Chapter 4. You have to know what you are intending to communicate and you need to know who your audience is. Without knowing the answers to these questions you will fail at the first hurdle.

### Your intention

You need to know why are you writing your report or developing your presentation in order to create the right tone throughout – for example, are you:

- presenting the findings to an incident investigation;
- demonstrating a business case;
- providing a progress update;
- explaining the position on something.

Next, you need to think about the 'so what?' question. This is what your audience will be asking themselves as they read your report or sit through your presentation. What do you want them to do? For example, do you want them to:

- endorse your suggestions as the way forward or change of policy;
- agree your findings;
- be more informed about the subject.

### Your audience

Who are you creating the report or presentation for? In other words, who is going to answer the 'so what?' question. As a general rule, the style of a board paper or presentation is completely different from doing the same for operational or front-line management. The only exception to this tends to be where you are part of the team you are presenting to.

It is important to do your research into your organisation's house style, ask for copies of previous good reports and presentations to get a flavour or just ask people what works round here. If you don't, you would not be the first person to assume that because you are presenting to the executive committee you should therefore be more clinical in your style, only to find that they like a more tactical style. While not a show-stopper in itself (you will remember for the next time), it just means you have to work harder to get going – and why make life more difficult for yourself?

## Writing an effective report

Armed with the information above, you can now start to think about the content of the report. Here we will not discuss the report's content as that would be a waste because they will all be different and after all as a qualified safety professional you should be comfortable with this element.

Experience shows that a good format to follow is simply:

- **Executive summary**  Always think of your audience; most people at work are busy people and many are sent information that they have to read but they do not necessarily have enough time to wade through pages and pages of a report to get to the key items. Therefore, starting with an executive summary is really helpful. Here you should summarise in no more than four paragraphs what the report is focusing on and the key findings. This will enable the reader to get a feel for the report as well as helping them decide if they need to delve deeper into it. It is really aimed at those you might copy in to the report, rather than the main recipients.
- **Introduction**  Here you should explain what the report is going to cover and the background to the subject as well as outlining the aims and objectives of the report.
- **Main body**  Depending on why you are writing your report (see above), this is where you put all your findings, thoughts, ideas or justification. There is nothing wrong with using subheadings or even numbered

paragraphs – in fact, they often help you create a logical argument or discussion and that is exactly what you are trying to create. There is nothing worse than reading a report that takes you all over the place; remember, a report is about sharing things in a structured way.

- **Recommendations**   Unless you are writing a positioning paper, in which case all you are doing is explaining where you are with something, the chances are you will be making some recommendations for people to consider, in which case it must be really clear to the reader as to where they link to the things you have discussed in the main body of the report. Depending on house style these can come immediately after the topic in the main report (usually highlighted in some way – e.g. through using italic text) or in a separate section afterwards.

  To help illustrate the point about the clear link between the main findings and the recommendations, imagine you are the safety professional carrying out an audit of a wood-working workshop and you have found missing guarding across a number of sites and you think the best way forward is for the engineering managers at each of the sites to do a complete review of all their guarding, you could write something like:

  > During the inspection a number of machines were found to have missing or otherwise defective guarding. While the Engineering Manager took these out of service at the time until they were made safe, it is recommended that a full review of machine guarding takes place across all locations to ensure that it is present and functioning correctly. It would also be worthy to review why the pre-use inspection checks failed to identify these problems.

- **Conclusion(s)**   This is where you draw together the various strands in your report and pose the 'so what?' question you have identified you want the audience to answer. For example, imagine a report outlining the business case for the deep clean on the floor in Chapter 8. You might conclude that paper with something like:

  > The risk of colleagues slipping on this floor remains and has caused 4 lost time accidents in the last 12 months. Taking the action outlined to prevent such injuries will not only improve safety but also deliver a cost saving worth approximately £174k over the next three years and therefore it is recommended that we take this option forward.

## Common mistakes

Experience shows that there are a series of common mistakes made in reports written by safety professionals which devalue the overall message. The good news is that they are easy to spot and relatively simple to fix:

- **One large paragraph**   You might laugh but you would be surprised how many people write in one large paragraph. A paragraph is a series of sentences that relate to a single topic, so remember to press the return button twice to start a new one when you are moving on to a different subject, irrespective how minor the difference.
- **Double space after a full stop**   Again, this might sound obvious but experience suggests that it isn't. This is about the basic rules of word processing and it makes the whole document more professional and easier to read.
- **Abbreviations**   Remember, do not use abbreviations unless you have already explained what they are the first time you have used them in the text. For example:

  > the use of personal protection equipment (PPE) is widespread to help further reduce the risk of personal injury and is used as a last resort . . . PPE is also used in combination with other control measures to help make them more reliable.

- **First or third person**   In most cases reports are best written in the third person rather than using 'I' as this makes it less about your opinions and more professional – e.g. it is recommended that . . ., consideration should be given to . . .
- **Avoid a thousand words**   Often in the safety professional's desire to get their message across they have the tendency to use a thousand words when ten or twenty will do. It is important to try to remember this and apply it wherever possible. It will endear you more to the audience and if they want more information, you can always provide it upon request.

## Writing and delivering an effective presentation

The first few steps for writing and delivering an effective presentation are the same as writing an effective presentation. Where things begin to differ, though, is the way in which you present the information. The default position for most people is that a presentation equals the use of slides as the medium used. While there is undoubtedly a place for slides there are other options to presenting as well which, depending on the circumstances, can be more effective. Before we explore some of these, let's look at how to make slide-based presentations really effective.

Often, presentations are used to support a report and this is normally circulated in advance. Where this is the case it enables you to put down the key points and focus on the areas you want the audience to focus on. Like a report, experience shows there is a general format that works rather well:

- **Title slide** – say, with a snappy title, what it is you are going to talk about.

- **Aim and objectives** – from the outset say to your audience what it is you are hoping to achieve with the presentation and how you will do it, what are you going to cover.
- **Main body and recommendations** – like a report concisely provide the detail needed on the subject to fulfil your aim.
- **Conclusion or summary** Again, like the report, conclude with the key points.

You should also bear in mind that you could include a slide that provides an executive summary just in case you don't have as much time as you might expect – after all, we have all been in meetings that over-run and 'air time' is cut.

Writing the presentation is only a small part of the story as how you deliver it is arguably more important. If you are ever unsure about how to deliver an effective presentation find out who the organisation's best sales person is and then go and watch them delivering a 'pitch' to a potential customer.

To help make your presentations really effective, there are some basic points to remember:

- **Stand up**   This is very important even if others in the meeting don't when they present. It creates a natural focal point for the audience and allows you to have eye contact with them. It says 'I'm here and I'm confident about what I'm talking about so listen up'.
- **Show some passion**   Speaking with passion and conviction is vital. If you come across as bored or even speak in a monotone voice, how can you expect your audience to be excited and fired up by what you are saying?
- **Tell a joke**   Admittedly, some people cannot tell jokes and if you are one of them do not try to. However, if you can, then do not be afraid to bring a little humour into the presentation. It will help break the ice and endear you to the audience. Remember, health and safety does not have to be boring.
- **Be confident**   The chances are that unless you are talking to a room full of other safety professionals, you will know more about the subject you are talking about than your audience, so be confident.
- **Don't hold your notes**   You do not need notes if you are using a slide presentation. The idea is that the slides have a little text on them and act as a prompt for you to talk around. In other words, have an idea about what you want to say and say it. It will be far more powerful and natural. By all means have key points to hand in case you get a question about some detail of what you are talking about, but only refer to them when necessary.
- **Don't read slides word for word**   Only do this if you want to lose your audience in seconds.

- **Tune into your audience**   Often overlooked, but if you can tune into the mood of your audience in relation to what you are saying you can detect which bits you can go through a bit a quicker and which bits you need to focus on. Think about it too from their perspective. There is nothing worse than a meeting overrunning and the presenter insisting on going through every slide. The chances are that the audience has already mentally packed up and are heading out of the door.
- **Be sparing with animations**   If you use too many animations you run the risk of giving the audience a headache or making them think that you cannot be very busy if you have so much time to spend showing so many slides.

## Other ways to present effectively

As Chapter 4 explains, there are many ways to communicate and there are numerous different ways to present, and while admittedly they do take you out of your comfort zone, taking a well-managed risk, in the right situation, can deliver outstanding results. Here are three alternatives to slides that experience shows do just that.

### Flip charts

As we have said before, these provide flexibility to the person using them. In this case they allow you to build up the presentation – in other words, you can write or draw the key points or discussion points as you go. They work better in situations where you are presenting via conversation and need to refer to drawings or similar occasionally.

### Slides by another method

An alternative to numerous slides with animation is to prepare some sticky notes with the key points on in advance (with the text large enough so that people can see it) and then place them on the flip chart in a logical order as you move through your presentation.

### A powerful conversation

The most powerful yet difficult way to present information certainly to a group sitting around a table is to sit among them and talk. This works well where you are simply talking through a paper or idea, provided you can project yourself such that everyone around the table feels as if you are talking to them. Experience shows that the best way to do this is to move your chair back a little and push your body forward so you are on the edge of the seat. Then talk with real passion, making sure you look round the table of people regularly, ideally making eye contact with everyone. You have to be confident (not arrogant) to do this style of presentation as it effectively puts you as a peer to those you are presenting to. Get it right and it will be a revelation to you; get it wrong and you will lose your audience – and remember people have long memories. If you decide to give this a go, make sure that you have researched what the etiquette is and that your slot has been positioned well; there is nothing worse than sitting down to talk something through when the audience expects a more formal style.

## Conclusion

Communicating through reports and presentations is often a 'must be able to do' on many job advertisements for safety roles. The importance of being able to do this is more than just about getting a job, though. Without clear and effective communication, you will fail to get buy in, traction and engagement.

Every organisation will have its own house style, the way it does things. While it is important to tap into this, it is also worth bearing in mind the more general ways to make both reports and presentations as effective as possible, remembering to put your own personal stamp on them.

# Chapter 10

# Managing conflict

In every walk of life you come into contact with conflict situations. The trick is how to avoid them in the first place while still getting a result that suits everyone, or if this is not possible, to navigate your way through the conflict to get the same result. Often safety professionals feel that they seem to find themselves in more conflict situations than other functions and to a point this view is probably true – after all, sometimes you do have to stop someone doing something they want to do because it is unsafe. However, experience would suggest that often the way to avoid conflict is down to how you approach the subject and that challenge is a universal one, not just one that affects safety professionals.

Conflict can occur in the board room, on the front line, in a safety committee meeting or even in managing your own team. In this chapter we will look to understand what conflict really is, how to avoid it, as well as how to manage the situation when you find yourself in the middle of it.

## What is conflict?

Conflict is a sustained argument between parties that is driven from opposing views which, if not resolved, can lead to poor organisational performance and culture as well as feelings, on a personal level, of unhappiness. Very rarely in a work context does conflict spill into violence.

There is a big difference, though, between an ongoing debate and conflict. A debate is often what we find ourselves having and we can sometimes misinterpret this as conflict. A debate is a healthy (if lively) discussion of the points which results in a resolution, although sometimes not all parties are happy with what that is.

Debates are a healthy state as:

- People engaged in them are usually interested in the subject and keen to find the right resolution for the organisation.
- They are useful to help work through the finer details to make sure the decision is workable.

- Many people find it easier to make sense of things when they vocalise them and talk them through.

## Avoiding conflict

The best way to manage conflict is simply to avoid it in the first place. There are two key skills to help you do this: learning how to pick your fights as well as how to fight them.

### Learning how to pick your fights

Try not to fall into the trap of 'arguing' everything. Some fights you are simply not going to win, but by the same token do not ignore them – highlight the problem in writing and chip away at it. Although some things you will have to go out on a limb for, particularly if someone is likely to be seriously injured or worse. Wherever possible, try to get out of your silo and think of the bigger picture as some issues will resolve themselves without you falling on your sword for them. For example, a safety professional identified in an audit that bunding was required around a number of tanks, although to do so would cost a great deal of money due to their location and associated pipe work. The rest of the audit highlighted serious issues with machinery safety that needed money to be spent in order to resolve them. Instead of making a stand on both points, knowing there were changes coming via environmental requirements for bunding in the next 18 months that the organisation would have to do anyway, they flagged the issue and focused on the machinery aspects. Both were done in the end, the business managed their risks effectively and there was no conflict.

There will be some people in your organisation who you know will argue everything you suggest. You can either get annoyed with them or try to find a way through it; finding a way through it helps you avoid conflict. Most people have sense checkers, people whose opinions they respect and listen to. You need to find out who they are and influence them. For example, a safety professional wanted to alter a process but knew that the operations director was likely to be opposed to it 'just because' and they were worried if they went direct to them they would get into a protracted argument about it which would result in an unhealthy situation. Instead, the safety professional knew that the operations director had two area managers who worked for them who they were close to and listened to their opinions because of their experience. The safety professional decided to meet the area managers separately to explain why the process change was necessary and work through the implications. Once they had got their buy in, the safety professional arranged to meet the operations director to explain it to them, mentioning that their two sense checkers were comfortable with it.

No decision was made at the time but shortly after, the safety professional received an email from the operations director saying the change was 'fine by her'. Admittedly, on the surface this approach might seem a little long-winded, but if you compare it to the alternative it is actually quicker and the conflict is removed from the situation completely.

### Learning how to fight

If you think about most of the conflict situations you have found yourself in, the chances are if you had stopped and thought about how you were going to approach the situation you would have probably done it differently. Take, for example, noticing someone working unsafely at height installing new flood lighting; clearly, you cannot ignore the situation but you can think about how you deal with it. Shouting to the person to get down and going at them 'hammer and tong' is most likely going to end up in an undesirable situation and is less likely to help them change their behaviour in the long run. Taking a different tack of: 'Hello mate, can you come down I need a word?' and then asking 'What are the risks with you working up there like that?' and having far more of a coaching conversation is going to help defuse the situation and encourage them to think differently. A valuable lesson to learn too is that there will only be one winner in a large meeting and that is going to be the senior person in the room, so to avoid any conflict (and potential loss of credibility) make sure you have done all the ground work before a meeting.

To illustrate this, let us use the operations director example above, except this time the safety professional takes the proposal to change the process to the operations leadership team meeting. They explain the reason why it needs to change and the implications on the operation. The operations director is clearly not comfortable with it and makes their thoughts known. The safety professional tries to convince them but does not get very far and the operations directors still says 'no'. There is nowhere for the safety professional to go now. Influencing outside such meetings as explained above is critical as well as remembering that in situations like the example above, the safety professional is unlikely to get things changed nor will the organisation get anywhere near the solution it needs.

## Managing conflict

So you have exhausted every route to avoid conflict, and you find yourself in the midst of it. To help navigate your way through such situations successfully there are two commonly used tools (scripting and stuck record) that are extremely effective and both require you to be assertive.

### Being assertive

Often people think they are being assertive when in actual fact they are just being aggressive. The difference between the two really comes down to why you are behaving in the way you are; aggression is normally based on you winning whereas assertiveness is about you being clear on what you want while considering the needs of others.

Being assertive is not easy, particularly if you find yourself in a situation with people in more senior positions than you in your organisation or people outside of it. The keys to success are:

- You are very self-confident. This does not mean that you are arrogant; it simply means you have confidence in your competence, which goes back, in part, to the need for you to have sound underpinning knowledge and relevant experiences.
- Remember that everyone is responsible for their own behaviour. If someone reacts to what you say aggressively or with resentment, you cannot help that, nor should you shy away from the subject to spare their feelings. If it is the right thing to say, you should say it, just make sure you say it in the right way.
- You are responsible for how you behave, so remain calm.
- You can disagree with others' views, but do so in a non-confrontational way.
- Accept other people's views even if they are not the same as yours.

### Scripting technique

This technique is known by different names but it is a very popular approach. It helps you to be assertive and successfully navigate your way through a conflict situation. It is a conversation that is split into four parts:

- What is happening to cause the issue.
- What are the implications of the issue or how it is making you feel.
- What you need to change or happen.
- What will happen if things don't change or happen.

This is best explained with an example. Let's imagine we are the safety professional who came across the person working unsafely at height we mentioned at the beginning of this chapter. They have come down off the ladder:

> Tom, I have to say what you were doing up there was really unsafe. You were stretching over to one side, which means you could have toppled off and seriously hurt yourself. I couldn't just ignore that, that's why I came over. I know it will add a bit of time to the job, but if you

need to reach across to get the cabling, you will either have to move the ladder or use the cherry picker. I'll let Gareth [Tom's manager] know what we have said and will be back in a minute to see how you are doing.

### Broken record technique

This is a commonly used assertiveness technique that can be used to help manage conflict and it works along the same lines of a vinyl record getting stuck and playing the same thing over and over again. The important thing with this technique is to keep repeating your message over and over again while acknowledging what the other person is saying and eventually they will get the message. There are two potential downsides:

- It can become very irritating for the person on the receiving end, which can mean that the situation escalates even further.
- Used wrongly, it appears that you are bullying someone into doing something they do not want to do.

To bring this to life let us use an example of a safety professional talking with a wood-working factory manager about introducing a mandatory rule to wear safety glasses while on the factory floor:

*Safety Professional (SP):*   We need to introduce a mandatory safety rule that everyone wears eye protection when they go onto the factory floor because there is a chance of saw dust or other debris going into their eyes.

*Factory Manager (FM):*   I'm not agreeing to that, it will cost us a fortune.

*SP:*   I appreciate there is a cost implication, but we have already had three people off this year with debris in their eyes and that has cost us a lot too, so we need everyone to wear eye protection when on the factory floor.

*FM:*   They will never wear it anyway and I'm not going to buy different styles for everyone, it's not a fashion show.

*SP:*   I agree we don't have to buy individual styles for everyone, but we do need everyone to wear eye protection when they are on the factory floor, so I'd suggest we get a few pairs in to see which ones they like and then go from there.

*FM:*   All right, but I'm not wearing them – I've just brought some new prescription glasses.

*SP:*   Trouble is, though, we can't have one rule for one and one for others. We need everyone to wear eye protection when they go on the factory floor, and if you wear the safety glasses they will get dirty rather than your new ones.

As the example above shows, there could come a point where the factory manager will not change his or her mind, in which case you need to do something else.

## No progress?

While it may not be the right thing to say, it is nevertheless true that sometimes despite your best efforts you cannot overcome the conflict or inaction. If the blockage is a middle manager or lower you can, of course, go over their head and go to their boss for support. However, what happens if that support is not forthcoming? You have identified the real problem, but what then? Get your boss involved but that may still not sort things out.

In the end you have two choices: either call time and move on or keep your head down and battle away. The first option is the easy way out for sure – just pity the next person who comes along. The second choice is not for the faint-hearted. If you do stay, however, remember to put things in writing because when things go wrong (and they will – it is normally just a matter of time), you need to be able to defend yourself as organisations that do not take health and safety seriously tend to use the safety professional as the 'whipping boy' far more than those more enlightened organisations.

## Conclusion

Avoiding and managing conflict is a useful life skill; it can help improve your home, work and social life. Avoiding conflict is a lot easier than you might think. It requires being more balanced about the situation in which you find yourself and asking yourself whether you are really just having a debate. The most effective safety professionals are those who manage to avoid conflict not because they are pushovers; far from it, what they actually do is come up with ways to get the right solution for the organisation by being smart in how they get people on board with it.

Sometimes, of course, this approach does not work, in which case you need some tools to help you tackle it. This is not about being aggressive or shouting louder than the other person; there is no place for that in the modern workplace let alone the safety profession. Success involves being assertive and knowing when to pull back and rethink your tactics. There does come a point, though, when you have done all you can and, as odd as this may sound to some, you need to move into self-protection mode.

# Chapter 11

# Problem solving and making effective decisions

It seems that every job advertisement for a safety role these days requires good problem solving. If it is such an important aspect of the role (and make no mistake it is), it is somewhat surprising that many of the safety qualifications we undertake to achieve professional competence do not have at least a section on it. Those completing safety courses in higher education will learn such techniques albeit applied to their learning rather than the subject matter, and other people just describe problem solving as something they do naturally.

In this chapter we will briefly explore the nature of problems as well as a number of problem-solving techniques that experience shows are helpful to follow to solve problems in a systematic and structured way, all of which is based on effective decision making.

## What is a problem?

A problem is a situation or circumstance that has two or more variables, which move independently of each other, where there is no immediately obvious way for them to work in harmony together to create a solution. The best way to illustrate what a problem is would be to think about the best-selling toy of all time having sold in excess of 350 million units worldwide. It was first invented in the mid 1970s by Professor Erno Rubik, to find different ways for the professor to explain spatial relationships to his students. Each face of the Rubik's Cube is made up of a further nine faces each with the same colour. Each face can turn independently, meaning that in a few spins you have mixed up all the colours, creating the problem of how you return it such that all the colours on each face match. Just to show how complex a problem it can be, the Rubik's Cube can be scrambled some 43 quintillion different ways (1).

Some problems are good, like the Rubik's Cube, as they keep your brain active and, while you might try to apply some logic most of the time, you try to solve them by trial and error. From a technical safety perspective, we know this is not ideal, as when people apply trial and error it can invariably end up with an accident or near-miss. The same applies to business or safety problems, if we apply trial and error; sometimes you get the answer right, sometimes you don't. The problem is when getting it wrong costs the organisation money, your credibility and other people's safety and health, you need something a little more solid than trial and error.

## Effective decision making

Problem solving is just another way of saying 'you need to make effective decisions'. To do this, you need to follow a decision-making structure, like this widely known and followed example:

- **Define**  The first step in problem solving is to be really clear what the problem is you are trying to solve. Break it down such that the scope is really clear.
- **What data do you need?**  Decide what information you need to help you understand the problem situation.
- **Information gathering**  Go and get the information you need.
- **What could you do?**  Based on the information you have gathered, try to work out what the various options might be to help you solve the problem.
- **Decide which option you are going to do**  Given all the feasible options you have determined, which one do you think is the best to solve the problem?
- **Implement the solution**  Put in place the solution you have decided upon.
- **Review**  Check to make sure that the solution is actually having the desired effect – i.e. solving the problem.

This model is based on the Plan–Act–Do–Review model that is well known to the safety profession, particularly as it is the basis of health and safety management systems (see Chapter 6). Repeated exposure to this approach will mean that you do it in your head automatically rather than running through each block as a discrete item.

## Problem-solving tools: the art of the possible

Accepting that the process above is sound, the trouble for us is that we tend not to have many tools to help us to decide on the art of the possible when it comes to solving problems. There are a number of tools that can be used to help identify solutions to problems, though, and here we will look at two of the more common ones.

## Brainstorming

This is a well-known technique that is often used in group settings to record people's spontaneous ideas to solve a problem. The idea of a brainstorming session is not to work out which idea is the one to go with, but to work out what the ideas are; in other words, it is about the art of the possible. While, of course, there is a structure to doing brainstorming sessions effectively, involving setting a time limit for the session so that you throw lots of effort into a concentrated timeframe, arguably the most important aspect is that no idea is a silly one.

## Reversal

This technique is similar to brainstorming except that instead of trying to solve the problem, you identify what you could do to make the problem worse. Ironically, this will help you identify what you can do to stop it. For example, if the problem you have to solve is that your lost time accidents involving forklift trucks and pedestrians are increasing month after month, you need to address it. Ask yourself if you wanted to make the situation worse, what would you do. In all probability you would come up with the following list:

- Introduce the idea that forklift trucks have the right of way over pedestrians.
- Increase the speed at which the trucks can travel from 7mph to 15mph.
- Do not rethink the location of, nor repair, nor remark the safe pedestrian walkways.
- Encourage the truck drivers not to sound their horns or use the convex mirrors when passing through a doorway.
- Leave in place unnecessarily complex safety rules that nobody follows.
- Reduce the amount of supervisory monitoring.

From this list you can then use the tools below to help you identify what you need to change to improve the situation and so stop things getting worse – for example, introduce speed limiters set at below 7mph or increase supervision.

## Making the effective decision

Having established a list of things you could do, you need a way to decide which is the right one, there are a number of tools that can help.

### Someone else's view

Having used one of the other tools to determine what could be done to solve the problem, you need to go about working out which is the one you will implement. Depending on the circumstance, you could try to pick the solution by putting yourself in someone else's shoes to see things as they would. For example, think about a safety professional you really respect. Then think about which solution would they go for. Some years ago a safety professional was facing an uphill battle to engage a relatively senior manager in their organisation to focus on achieving compliance with an improvement notice. Having exhausted all the obvious options to overcome this, they put themselves in their old boss's shoes and remembered they would have gone above the manager's head and said 'It's my job to protect you and to do that we need to work together'. So they went to the functional director, explained the problem and asked 'How can we get them to focus because if we don't we will both be in trouble?' – a simple example but it illustrates the point. You can, of course, think of things from other people's perspectives like the regulator, shareholders, your family and people in the business community. Each will give you a different take.

### Biggest bang for your buck

There is nothing like money to help you think straight, imagine you have the range of possible solutions laid out in front of you. You have your last £100 in the world, which option would you bet your money on? A variation of this is the £100 bet. Instead of metaphorically betting all your money on one solution, imagine you split your last £100 between two or three options. The way you spread your money would help decide which solution to go with.

### SWOT analysis

In Chapter 3 we used a SWOT analysis to help present the situation analysis in a safety strategy. In essence what a SWOT analysis is, is a summary of what is good, weak, possible and really bad about an idea. Carrying out a SWOT analysis for each possible solution you have identified helps you to compare and contrast them with a view to determine the right one to go with.

## Conclusion

Problem solving and making effective decisions is not difficult for simple issues. The more moving pieces you have, the more difficult it is to come up with the right solution. Success comes from understanding what the problem is in the first place and researching it so that you are acting in a more objective way rather than going on a hunch. There are a range of tools to help determine what the possible different solutions are to a problem as well

# Chapter 12

# Managing change

Change is a constant thing in the modern workplace and experience suggests that at times safety professionals are a little reluctant to embrace it. Taking a different attitude to change helps enhance your credibility and getting involved in the change management process enables you to show more of your skills, which will help demonstrate the value you can add to your organisation.

In this chapter we will not be looking at how to develop, introduce and manage a safety change programme – there are plenty of other books on change management. Instead, we will focus on the role that safety professionals can play in the process as well as the opportunity to have a more open mind towards change, which will not only improve your mental well-being but, if exploited effectively, can help improve the organisation's safety performance and culture to boot.

## The implications of change on safety

Arguably one of the biggest risks facing safety professionals these days is that of managing safety during periods of change – and it is something that few of us cover in our qualifications we take to become safety professionals. Organisations have always changed in order to maximise efficiencies in delivering greater returns to shareholders and other interested parties. In times of a steady economy, or even boom, this tends to happen from experience every three years or so. In recessions, though, it seems to happen more regularly and understandably so, as boards across the world try to safeguard their organisation's financial future.

One of the negative things that change brings with it is uncertainty and that is why it is an issue for safety. Imagine the following two scenarios that are probably happening right now in thousands of companies around the globe:

- You are a manager who, through a restructuring process, now works in a new team made up of some people you have never worked with before or only know by reputation and with a new director who has a different operating style to your old boss. It's only natural that you feel under pressure to perform and deliver the 'result' (whatever that is) and, of course, you're worried about whether you fit in the new world.

- You work in a production plant that is likely to be moth-balled with production moving to another site 20 minutes down the road. You need the job to pay the bills and look after your family. The problem is that there are only 20 new jobs at the other plant and there are 50 people at the one being moth-balled who are likely to apply. Your mind is bound to be preoccupied with the future and the chances are that you are going to be feeling more than a little disengaged with the organisation for putting you through this.

Hopefully, you can see how periods of organisational change bring with them a potential loss of safety focus, which in turn can create a very real threat to the organisation's safety performance and culture. Yet such an effect can happen with any sort of change even when that might not be viewed as a 'major' organisational one. Take the changing of a shift pattern in a call centre or the installation of a new piece of kit that fundamentally changes the way things are done. The feelings of uncertainty in the people affected by the change will still be there – e.g. will I be able to operate the new equipment? If the kit is that efficient, will they need me? And so, too, will the challenges associated with the kit have an effect on the team's safety culture and performance?

## Change management model

Although we are not looking here at how to develop, introduce and execute a safety change programme, it is still important for us to consider the process of change. There are a number of steps that are commonly accepted as being key parts of a change management process, which at the basic level are:

- **What is the 'burning platform'?**   Here you need to define why the change is needed – it needs to be a compelling reason for people and organisations to get behind it.
- **Design the programme**   What are the elements of the change and what are the expected costs and benefits (is there a strong business case)?
- **Plan the delivery**   How do you intend to make them happen? Who is needed to help and when do you expect things to happen?
- **Establish a sound communications plan**   Culturally, change is often won or lost with the communications that are given throughout the programme; it is essential that a good communications plan is defined.
- **Get the right people on board**   To deliver anything you need the right people to help, which applies to change management too. This does not

mean that they need to be working on the programme full time (although depending on what it is, this might be necessary), but you certainly need relevant stakeholders to be able to support it with their time and resource. Here it is important to get a senior level sponsor to help remove any organisational blockers there might be and ensure that the resource is aligned appropriately.

- **Get the mandate**   No matter what the change, there will be some sign-off process that needs to be followed. This might just be your boss or you might need to get board sign-off, depending on what the programme is, but irrespective you need the mandate for change based on the work you have done thus far.
- **Delivery**   Now the hard work begins as you put into place the change programme.

Although rudimentary, these stages apply to all types of change from big corporate shake-ups to the introduction of new working patterns. Arguably from a safety point of view, provided the changes are well thought out and the implications assessed with the relevant controls in place, coupled with clear communication and increased proactive monitoring of safety during the change period, all should be well.

## Opportunities

Change presents opportunity. Although such opportunities might not be immediately obvious to you, there will be some. To help illustrate this we will explore three such opportunities arising from different change management programmes.

### Validating an organisational change

An organisation was restructuring their workforce and was making redundant several roles, transferring some of their responsibilities to others and broadening out other roles. The organisation's safety professional was asked to help bring some rigour to the process and help the organisation ensure that they were not increasing their risk exposure through completing the change.

To do this, based on the rationale for the change and the revised organisational structure, they undertook risk assessments with the relevant line manager for each of the roles they were changing. They identified the critical aspects of the role (the hazards) and determined the likely impact and likelihood of removing them to get the risk level. From this the relevant line manager was then able to work out how they would either eliminate or otherwise control the risks associated with each. For example, one of the roles played an important part in linking up with the organisation's trade body. Not doing so would mean that they would not necessarily know what was going on in the wider industry and this could have a detrimental effect on their competitor intelligence and industry best practice. To mitigate this risk, the organisation decided to get their functional experts to 'face off' with the relevant parts of the trade body – i.e. the safety professional would go to the safety committee. While increasing the work for those individuals, there were additional (and obvious) benefits to the organisation and the net effect on people's time was broadly neutral as they had to prepare and debrief the other person when they went to the meetings in any case. There were other examples where it was decided that the aspect of the roles presented little or no risk to the business, so no further action was needed.

To record all these changes in a structured and systematic way for future reference and good governance, disposition statements were developed. These were simply two columns for each affected role. The first column listed the items from the job description of the proposed redundant role and the other listed who was going to pick up that responsibility or whether it was no longer needed. Having done this, the safety professional prepared 'go live' criteria, which outlined the critical things that had to be in place before the change could happen and, if they could not be achieved, the relevant manager had to put in place a plan to bridge the gap with a definite timescale.

### A new team

We have all been in this situation: there has been some sort of change in the organisation and you find yourself in a new team reporting to a new boss. You want to be accepted by the others quickly and show your new boss your true value. When this happens in an operational setting, experience shows this is the best time for the safety professional to get involved with this team to help establish the right direction for safety. This is because they are less likely to want to speak out too much against anything put to them as they are trying to figure everybody out, including the boss.

To explain how to make the most of this situation it is easier to use an example to illustrate the point. Imagine you are a safety professional who has been supporting an area manager for a while and their area has just been reviewed so that they are now looking after a number of locations which are new to 'the team'. Having spent some time with the area manager to take them through what it is you want their location managers to do (it might be that you want to make sure that all their locations are up to the same level of safety the area manager expects, and to do that you will need the location managers to do A, B and C). Once you have their buy in, you attend their first meeting with the new team. You have already asked the area manager to open your session with some words saying how important safety is, you do your bit and the area manager finishes off by saying 'I want you to do what they [you] have just said'.

From experience you will see action from those in the team and what you have just achieved is something very smart.

- You have got everyone in that team to focus on safety even when they are uncertain about the future – after, they will not want to be the one who lets the side down.
- You have set the tone for all future safety interactions with the team: 'my boss thinks this is really important' which, as we all know, is the first step in getting any safety culture right.
- You have showed the group how closely aligned you and their boss are; this will help build your credibility with people in the team you haven't come across before and it will also come in useful as people start to challenge you as the team develops.

As the team develops, a pecking order of location managers will develop and for the safety professional it is important to note this order as there are more opportunities for you to focus people's attention on safety and positively affect the culture and performance of the team during this period of change. It is normally pretty apparent what the pecking order in the team is going to be and spending time with those people who are going to be more dominant is time well spent. The word 'dominant' conjures up all sorts of images, but in this context they are the sorts of people who others in the team will listen to, respect, and the boss will use as a sense checker of ideas and proposals put to them.

### Supporting a change programme

As we said at the start of this section, change presents an opportunity you just might need to find if it does not leap out at you from day one. The following example illustrates this further. A service organisation was introducing a programme not dissimilar to lean manufacturing techniques with a view to streamlining workshop activities and help reduce the cost base. At the kick-off meeting, it was apparent that the programme manager had got senior level support for the programme and it was not going to go away. All the stakeholders in the room (including the safety professional) were unsure of the benefits the programme would deliver. To give it a chance, though, the safety professional took some time to talk with the programme manager over the coming weeks and realised that in applying the lean manufacturing techniques they would improve the housekeeping issues in the workshops and help review the organisational safe systems of work in the light of the process changes they were planning – this review was something the safety professional was planning anyway but was struggling with time to do so. As a result, they decided to get more involved in the programme and agreed with the programme manager that while they did not have the time to review the safe systems of work on their own, the programme would do so but the safety professional would be part of the sign-off process and if they were not happy they would not be published. The programme did not go away as some of the other stakeholders had hoped. In the end they came to the party although the opportunity for them to advance their functional agendas was reduced.

## Conclusion

Change is always going to happen. You can either accept and embrace it or resist it. Resisting change is a risky strategy as most of the time it will go ahead. There are a number of ways for safety professionals to show how useful their skills are in the change management process as well as ways to piggyback on wider change programmes to improve their organisation's safety culture and performance. It all adds to helping you become a more well-rounded professional who is versatile – qualities all employers look for.

# Chapter 13

# Managing projects

Many safety programmes fail to improve safety culture and performance, not because they are bad ideas, but because they are not developed and implemented properly. To help overcome this you need to use good project management techniques and these are useful irrespective of the size of programme or project you are managing, from the development and roll-out of a major change programme to simply delivering your personal objectives.

While there are globally recognised qualifications and standards relating to project management, in all but a few cases the safety professional is unlikely to need them. With this in mind, this chapter will explore how to manage projects simply and effectively, ensuring that the process you use for their development and implementation phases does not reduce the effectiveness of the result.

## What is project management?

A good starting point to answer this question is to try to explain what a project is. By its very nature, a project is temporary with a definitive start and end date, often with a set budget assigned to it and there is an expected set of outcomes. Projects can be 'run' by a single person or many people; they can be in a dedicated project team or do it as another activity aside from their day job. There is, of course, nothing wrong with any of these approaches; the key is making sure that they are right to suit your project and the terms of reference you have been set. Managing this activity, therefore, is about making sure it is delivered effectively with sound management through all aspects of the programme, including its planning, development, implementation and evaluation of the outcomes, as well as ensuring good control so that you hit the target.

## Project management steps

As we have already established, we are not going to explore project management in such depth that you can consider yourself to be a proficient project manager. However, in order to use some of the principles of project management in an effective way so that you get the most benefit from your safety programmes, it is important to have a basic appreciation of the various elements that make it up. There are a number of project management models you can follow, all of which are good in their own way for the right situation. However, they all have common threads.

## Planning

As we have already explored in previous chapters, it is essential to set yourself up for success and managing projects is no different. The first stage is to understand the project. Here you define what the project is, often producing something similar to a terms of reference. This includes questions such as:

- Background on the project: why it is needed?
- The scope of the project: where are the boundaries?
- What does the project have to deliver: what are the outcomes?
- What resources, including budget, are available: how will the project be achieved?
- Who is going to sponsor it and who will manage it: who is accountable for delivery?
- What are the expected benefits: what will you get for your money at the end?

## Getting project approval

This part is like asking which one comes first, the chicken or the egg. Do you do the planning first, then go with a completed business case for approval, or seek approval then develop the terms of reference and associated business case? Irrespective of this, there is a stage early on where you will need to get the 'OK' to do the project. Depending on what this is, you might need to prepare a business case, which is something we have already discussed in Chapter 8. Part of this process is about making sure the necessary resources are aligned to the project and not just the budget. Time is a precious commodity, so it is important that you have the right support in order to get your project team-mates' time freed up.

## Development

In this phase you develop the specifics of the project. Typically, this covers:

- the action plan, who is going to do what, when and how;
- the risk register;
- the communications plan.

The action plan hopefully is a self-explanatory element; here you decide the actual component parts of the project. They are often displayed in a 'Gant' chart, which plots them against the project timescales. This approach allows you to see all the different elements of the projects and the interdependencies between them. A risk register is often overlooked but experience shows that it is a vital aspect of ensuring an effective delivery. Much like the classic health and safety risk assessment we understand as safety professionals, the register here records all the things that could prevent or delay the project from delivering or having the desired effect. Once you have identified these, you can determine their risk rating and the necessary 'treatment' or, as we might call them, control measures. The outcome of this assessment feeds back into the action plan.

Having a clear communications plan is vital. Here you need to think about who you need to communicate with. It is often useful to think of people in three groups: those who will be affected by the project – e.g. it might change the way they work – those who will have an interest in it – e.g. they might be responsible for a function that interacts with the project's outcome and then everyone else who might like to know about the project but they just as easily might not. Once you are at this stage you can apply the techniques we have already discussed in Chapter 4.

### Implementation

In this phase you put the action plan into play while monitoring progress tightly to make sure there are no risks to the project's timely delivery. You also put your communications plan into place and keep the different stakeholders informed.

### Post-implementation review

A programme following a sound project management process will carry out an evaluation that looks at the actual things the project delivered as well as the way the project was managed, making recommendations for improvement for next time. Post-implementation reviews tend to consider whether:

- the project was delivered on budget;
- all the benefits have been achieved;
- there have been any unforeseen benefits as well as unintended consequences;
- there are any things that need to be finished off or could be improved upon.

This is all in addition to the obvious questions on what went well and what could be improved. Depending on what the project is, you might do this one week post-implementation only, or three, six and twelve months post-implementation.

## Real-life safety examples

Now we have briefly explored the theory of project management, we will see how it can be put into practice in two real-life situations.

### *New online incident reporting system project*

A safety professional was tasked with designing and implementing a new online incident reporting system to replace an existing ageing one. Other than that basic remit, little more direction was given. They set about developing a 'straw model' of what they thought a new system needed to look like and then spent some time discussing and refining this with various stakeholders in the organisation. From this they built clear terms of reference and defined the scope. Their sponsor helped them get the necessary support from various other organisational functions and when they were up to speed they met with several external software developers. Once they had agreed the price, they gained the relevant approvals.

The safety professional locked the scope down early on to counter scope creak, by which people keep adding to what the project is going to deliver so it becomes a big beast to manage and ultimately fails to deliver. In this instance as the project was to develop from scratch a ground-breaking multi-channel incident reporting system, the detailed plan of what was going to be built was created by the developer with the safety professional's timescale in mind. The working group then developed a risk register which included the foreseeable risks and mitigations, examples included:

- **Cause**   The vendors are unable to deliver the system in the time given, creating a risk that the target launch date is not achievable. The effect of this will be that the system is not delivered on time. Controls in place to manage this risk are:

  - initial tender conversations in which the vendors indicate that the target deadline is achievable;
  - detailed delivery plan with penalty clauses developed;
  - support provided for back office development;
  - manage vendor performance on a weekly basis.

- **Cause**   Stakeholders are not engaged with the system, creating a risk that the system is not adopted or viewed as an improvement. The effect of this will be that the full benefits of the system are not achieved. Controls in place to manage this risk are:

- develop stakeholder engagement presentation;
- provide further engagement session when the contract is let with a detailed delivery plan;
- provide fortnightly updates on progress, issues, risks and milestones for cascade to stakeholders.

As part of the overall control mechanism to make sure the project was on track, the developer and one of the working group had a weekly conference call to review progress, work through any blockers and make sure the developer was getting all the information they needed from the organisation to develop the tool. Every two weeks the whole working group and the developer had a conference call to review progress, agree what would be delivered in the next two weeks and provide additional direction to the developer as necessary.

As part of the communications plan, the safety professional provided a concise update to the main stakeholders that covered:

- progress made in the last two weeks;
- key objectives for the next two weeks, which included critical work that needed to be included other than the software's development such as getting the contract signed, starting to engage the wider organisation about the changes and what they would mean;
- help needed;
- a confidence score in the right time delivery of the project.

After the system went live, the work group completed a snagging list and undertook three- and six-month post-implementation reviews, out of which they identified several aspects of the system's back office that needed to be amended. The question here, then, is: would the project have been delivered without the use of project management principles? The answer is probably not, as everyone in the working group had busy day jobs to continue and did not have the technical software programming knowledge to really understand what the developer was doing. While not strict project management, without a clear and tight direction from the outset, without the right people being involved, a solid risk register and a clear communications plan, the chances are that it would have still delivered, but nowhere near the right timescale.

## Personal objectives

As happens in most organisations, a safety professional was set a series of personal objectives linked to their annual bonus. Not only did the safety professional want the bonus, but they wanted to change their manager's opinion that safety professionals cannot deliver tangible outcomes.

Working with their manager, they tighten up the objectives to note what 'good result', 'exceeded expectations' and 'outstanding' would look like for each of their objectives. They then spend half a day planning the discrete tasks for each objective, how they linked together and who they would need to engage with in order to achieve them. From this they created a mini-risk register which essentially said that if you spot any of these signs, you need to stop and find a way round it as you are about to hit a problem.

Then, at each one of their monthly one-to-ones with their line manager, they explained their progress with each objective, what their next steps were as well as areas they might need some help or guidance with from their manager. Their end of year performance review was the quickest the manager had ever had with a team member, as they were well aware of their progress and how they went about achieving what they did. Not only that, but by really defining what the outcomes were at differing levels of success, they managed to prove to their manager that safety professionals can deliver tangible outcomes.

## Conclusion

There are a number of different approaches to project management. However, all have common threads based around the Plan–Do–Act–Review model discussed in Chapter 6. Using the basic principles of project management can help you make sure that your safety improvement programmes have the desired effect, in the right timescale and within budget. Doing so helps build your credibility and helps you to be more in control of things and able to cope with a bigger workload.

# Chapter 14

# Using your time efficiently

Safety professionals always seem to perceive themselves as being under resourced. Whether this is a perception or a reality is immaterial, what is important is that you deal with things in the most effective way, which includes how much time you focus on things. Most people think that time management is 'obvious', yet it cannot be that obvious. Stop and have a look around you. The chances are there will be people making really bad use of their time – perhaps they are spending too much time on the wrong things, not enough time on the things that will make a real difference, or never getting anything done because their 'to do' list just keeps getting bigger.

Becoming an effective manager of time will not only enable you to achieve more; you will use less effort to do so, improve your resilience and, depending on the tools you use, as a safety professional you will develop your strategic thinking. In this chapter we will explore a number of time management tools that will help you to do all this and can enable you to create time so that you become really effective in your role and drive safety performance and culture forward the most.

## What is time management?

Maybe unsurprisingly time management is about making the most effective use of your time so that you achieve what you set out to do. To do this you need to be clear about what you want to achieve and what are the most efficient ways of doing so.

### Time efficiency

Experience repeatedly shows that for safety professionals there are six tools and approaches to managing your time that enable you to exceed your objectives, drive you to think more holistically and enable you to achieve a good work–life balance.

### Lists

The most basic of time management tools is to create a list. Indeed, many people do it, whether they are writing a list of things they have to do when they are off for the week or the tasks they need to complete for the month-end at work. You now need to put some prioritisation against the things you have listed.

The key thing is to make the next step and determine the priority of the things on that list as this will help you plan your time more effectively. To do so is not that hard or time-consuming. A common approach used by many is to simply take your list and put each item into a grid like that shown in Figure 14.1.

Now simply work through the tasks as they appear. There are two ways in which you can use this approach depending on your outlook:

- If you have many things to do, complete a number of quick wins, making you feel that you have achieved a lot; this will help you feel less pressured by the fact you have a large to do list, enabling you to focus on the harder, more time-consuming tasks.
- You might want to spend an hour a day on each of the harder, more time-consuming tasks so that you make some progress on them regularly as well as doing some of the easier tasks.

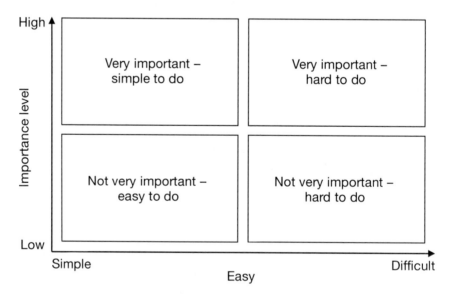

Figure 14.1 A simple time management tool

In this day and age, though, there are, of course, a number of software programs available on the Internet or for your portable device that will help you to do all of this. The good thing about this is that it does not matter how you like to work – hard copy or soft – you can use it to help you maximise your time.

## Being efficient by managing differently

Experience shows that the more traditional safety team set-ups tend not to lend themselves to the efficient use of time or resources, and it is common for such teams to be operating with 'silo mentality'. What this means is that individuals or teams are too focused on their specific job and are failing to acknowledge or take into account the wider perspective of the organisation.

A way round this is to be more efficient with both your time and resources to reconfigure the roles and responsibilities of the safety team while maintaining the headcount but freeing up more time to focus on the right things (which arguably is the work that will really change the organisation's safety performance and culture). The key to success here is to get each member of the team to work at the right level or, if the team is just you, to make sure you are working on the right things. Typically, this means moving the more trivial aspects of the job away from the safety professional on the ground to a central co-ordinator who is at the end of the phone or email inbox. Experience shows that ideal candidates for this are the safety team's administrative support (following some formal lower level safety training).

To test to see if this sort of approach might work, list all the telephone calls and emails you get over a one-week period. Then, against each entry put a tick if you really needed to deal with the issue and a cross if you didn't and somebody else could have done so for you – for example, calls or emails asking 'Can you send me . . .?' or 'What do I do about . . .?' Where the information is available in your risk assessment manual, this would warrant a cross, but things like 'I've looked at the risk assessment manual and I'm still not sure what to do, can you come and help?' would demand a tick.

If you are working less efficiently than you might, you will have numerous crosses on your list. Setting up the approach outlined above means people requesting the information from a central point rather than the safety professional. As a result, they can get the information much quicker as the person on the road would have to either remember the request or pull over if driving, write it down and then action it when they have the time.

Some contacts might be 'How do you do. . .?' or 'How should I deal with . . .?', in which case the co-ordinator can either help over the phone or arrange for the most appropriate member of the front-line safety team to visit the site to support the manager face-to-face. In effect, this approach puts in place a filter that captures around 90 per cent of all contacts that would otherwise go to the team and detract them from other things. From experience, this approach has the potential to increase job satisfaction at all levels of the team and the amount of contacts that can be filtered could increase if a 'knowledge bank' is developed of questions or problems and their answers or solutions.

### Managing customer expectations

Often being inefficient with your time is simply a product of you not managing your customer's expectations particularly well – customers being anyone for whom you provide a safety service. While this sounds like a criticism, it is easy to see why it happens; experience repeatedly shows that safety professionals want to help (it is probably one of the reasons they became involved in the profession in the first place) and as a result saying 'no' is something that does not come easily to them.

By managing your customer's expectations, you can become more time efficient and still keep them happy. Doing so comes down to three things:

- understanding them and their needs;
- ensuring you have the relevant products and services that the customer needs;
- the ability to deliver the products and services professionally in a timely and cost-effective manner.

Other chapters discuss how to work out customer needs and how to develop products and services that are 'fit for purpose', but once you have done so, what will happen then? How do you ensure that you have the ability to deliver what they want, when they want it? Typically, there are three choices:

- increase the headcount of the safety team;
- change the ways of working to free up some capacity for the team;
- develop service level agreements (SLAs).

The first option is unlikely to happen, at least not immediately. Freeing up capacity (as discussed previously) with a combination of developing SLAs is the most sensible way forward. Of course, if the SLAs are not good enough for the customer, with them supporting your case of more 'heads' in the team, they could potentially carry more weight than your lone voice.

In a nutshell, SLAs simply define what you will do and when. Typical health and safety related SLAs include:

- 'all lost time accidents will be investigated by the safety professional, this process will commence within one working day of the incident being reported';

- 'where five or more working days advance notice is given of a visit by an enforcing authority a safety professional will attend the visit to support the local manager'.

Before entering into negotiations about the timings in the SLAs, it is important to work out the likely demand on your time, because without it you could quite easily be setting yourself up to fail. The worked example in Figure 14.2 shows one of the simplest ways to determine your ability to meet their demands.

The example shown in Figure 14.2 is only an estimate of the safety professional's 'supply and demand' based on previous experience. However, it

| Work plan forecast – demand | | | |
|---|---|---|---|
| Product or service | Qty | Delivery time per unit (approx. in days) | Time |
| Lost time accident investigation | 12 | 1.5 | 18 |
| Enforcing authority visit support | 8 | 2 | 16 |
| Site audits | 40 | 1.5 | 60 |
| Managers training | 16 | 1.5 | 24 |
| Safety committee support | 4 | 1 | 4 |
| Managers meetings | 16 | 1 | 16 |
| Insurance claims support | 6 | 1 | 6 |
| Total demand: | | | 144 |

| Work plan forecast – supply | |
|---|---|
| Safety professional | Days |
| Holiday entitlement | 25 |
| Bank holidays | 8 |
| Estimated sickness | 5 |
| Continuing professional development | 5 |
| Total days unavailable: | 43 |
| Annual workable days: | 260 |
| Total supply: | 217 |

| Work plan forecast  – summary | |
|---|---|
| Demand: | 144 days |
| Supply: | 217 days |
| Difference: | 73 days additional days available |

Figure 14.2  A worked example of a supply and demand forecast

does show that there is some scope for them to deliver safety improvement programmes in the 73 'surplus' days. Developing SLAs and working out the supply and demand allows you to quantify the work you do. There is an added advantage to setting SLAs as they provide other performance metrics of how you are performing other than just the organisation's safety performance.

### Risk profiling

Imagine you have a number of different companies in your organisation, or a large number of sites in your region, or many different teams in your factory. Traditionally, the commonest ways to decide where you spend your time is normally based on who shouts the loudest and where the 'fires' are that you have to put out. Clearly, this approach is neither efficient nor clever. Luckily, though, we are in the business of managing risk and we can apply this thinking to how we manage our time and work efficiently. However, if you are working inefficiently, the chances are you do not have the ability to identify easily which parts of our organisation, region or factory need your support the most.

To help overcome this you can develop a simple risk profiling tool. Using a range of objective and subjective measures you can create a rating which can be aligned to a risk level – either high, medium or low. In Tables 14.1 and 14.2 there are some typical objective and subjective measures used in risk profiling tools.

Of course, these measures are just safety related. However, to improve the validity of the profile you could start to factor in other wider business measures such as:

- employee engagement scores;
- employee churn rates;
- number of customer complaints.

This approach allows you to list all your businesses, sites or teams to help you focus your time in the areas that will make the biggest difference whatever your driver is and, as important, you can justify why you are taking this approach.

### Delegation

Delegating tasks to people can help manage your time more efficiently provided you do it correctly. Experience shows that safety professionals like to keep a tight grip on things and there is nothing wrong with this provided that these are right. If they are wrong, you fast become time poor and fail to get others to take ownership for safety.

*Table 14.1* Common objective measures used in risk profiling tools

| Measure | Scoring |
| --- | --- |
| No. of RIDDORS | Actual number |
| No. of near misses | +10 if below organisational average |
| No. of employer liability claims | Actual number |
| No. of improvement notices received | 1 improvement notice = 10 points |
| No. of prohibition notices received | 1 prohibition notice = 25 points |
| Previous audit score | 100% – audit score: e.g. 100% – 82% score = 18 points |

*Table 14.2* Common subjective measures used in risk profiling tools

| Measure | Scoring |
| --- | --- |
| Regional manager's score | Out of 100 given by the regional manager based on their view of how well the site is managed *in general* (100 being bad, 0 being perfect) |
| Safety professional's score | Out of 100 given by the safety professional based on their view of how well the site *manages safety* (100 being bad, 0 being perfect) |

True delegation is about providing people with the opportunity to develop themselves, so you give them a task or project to let them develop their knowledge, skills and abilities. In the very short term doing this takes a little more of your time as you need to help get people up to speed, but in the long term you free up more of your time and the other person becomes more rounded and proficient. Once you are at this stage, all you need to do is monitor what they are doing to make sure it is right.

Here are two examples to help bring this to life:

• As part of a programme to improve product storage, an organisation wanted to find out what the products were that their front-line colleagues really did not like handling or were not too comfortable putting away. The safety professional wanted to do a survey of people but was struggling to commit enough time to getting round to all the team meetings at the factory to ask. Instead, they identified that there were two new safety representatives who could do this and use it as an opportunity to develop their presentation skills (and be seen to be working collaboratively with the organisation to improve safety). The safety professional squared their release with their line managers and briefed them accordingly. They sat in on the first session and gave them some feedback. Pleased with what they had seen, the safety professional let the safety representatives carry on with the rest of the sessions and sought feedback from the team's managers to make sure that they were still doing a good job.

- A serious accident occurred one day that necessitated it to be reported to the enforcing authorities immediately. However, the safety professional was already snowed under and then had to go to the site to investigate what had happened. He talked to one of the general administrators about how to make the report for them. When they saw what they had done, they were impressed and asked their manager if they could do all the reports, as they only took about an hour a month all together, which they agreed to. Now the safety professional reviews the report when they are sent electronic confirmation of receipt and provides feedback to the administrator for the next time.

### Programme priority tool

In Chapter 3 we discussed how you can work out how to prioritise which safety improvement programmes should be focused on over others. Although we discussed that context in relation to setting a safety strategy and work plan, it works just as well here, helping you to be more efficient.

## General tips

Up to now we have explored some tools and approaches that dramatically change the way you work to be more efficient. Experience shows that these will have a positive effect. However, in all but one case they require a great deal of planning and work to get them right. But what are the things you can do right now to help you to be more time efficient? There are six general tips that can make an immediate difference:

- **Use voicemail**  Do not feel compelled to answer your phone. If you are lucky enough to have a personal assistant, they will filter your calls so that you only take the ones you want. You can do this yourself if you are busy. Unless the caller is your manager or related to what you are working on, let it go to voicemail. You can then listen to the message and decide to respond at an appropriate time to suit you.
- **Work when you are most effective**  Some people work better first thing in the morning, others late at night. Get to know yourself and when you are at your most effective when you have a lot of work on. For example,

you might decide to do your project work in the morning and visit sites in the afternoon, if you are more productive early in the day.

- **Surroundings**  Just like getting to know when you work most effectively, find out where you work effectively. While you may not be able to work there permanently, at least you know when you are up against it if you have somewhere to go to get back 'on top of the job'. For example, people have mixed thoughts about working from home, but if you can lock yourself away, get things done when you need to and your manager is happy for you to do this, why wouldn't you? The chances are you will put more hours in and be more productive. In the same way if you are in a busy office, you might just need to book yourself a meeting room for an hour or two to get something done in peace.
- **Breaks**  When you have lots to do the tendency is not to take a break. Next time you find yourself in this situation, have a break – make a drink, talk to someone, but stop doing whatever you are doing for a few minutes, ideally taking in new surroundings, then go back to your work. You will be surprised just how refreshed you are.
- **Say 'no'**  Sometimes you just have to say 'No, sorry, I can't do that' or 'I can't do that in the timescales you need, sorry'. When you do this, how and to whom is really important, but there does come a time when you have to say no. Often we get dragged into all sorts of things that we do not necessarily need to be involved in or to the level of detail that we are.
- **Head down**  Every now and then, despite your best time management efforts, you just have to get your head down and get on with it. It is a mindset that you need to have and use as sometimes it is down to you – at which point it is important you know the right surroundings and time of day for you to be the most productive and get it done.

## Conclusion

Time management is a vital element of any role. Irrespective of the state of the economy, safety professionals feel as though they need more time, people and resources. This may be true and making a case for change is important, but it is also important not to get so wrapped up in it that you become a victim, as there are plenty of ways for you to become efficient in your ways of working and still help the organisation's delivery, improving safety performance and cultural change.

# Chapter 15

# Coaching

Here are two ways of tackling an unsafe condition or behaviour: 1) you take control of the situation telling people and their manager how they should be doing the job; 2) you stop the job and explore with the people and their manager how they could do the job safer. The first way is more akin to the fist-banging safety professional mentality that we have already said is less desirable, although in situations of imminent danger such a direct approach can be justified. The second approach is more likely to have a longer lasting impact and change people's behaviour, as they have come up with the answer for themselves, worked through the barriers that prevent them from doing it and implemented it. At a very basic level, the second approach outlined is coaching and this is the subject we will explore in this chapter. Make no doubt about it, coaching is a style we all need to be able to use expertly, when the circumstances dictate.

## What is coaching?

Coaching is a general leadership and development style that has become more and more fashionable in recent years. Fashion aside from a safety perspective, taking such an approach helps you move your organisation's safety culture from 'I will follow the rules, else I'll get disciplined' to 'I will work safely because it makes sense to me'. In short, coaching helps a person improve their performance or knowledge through discovering the answers themselves with the coach facilitating that learning rather than telling them what to do. Typically, it is based around real-life, hands-on workplace problems.

## A coaching technique

There are a number of coaching models but they are commonly based around four questions:

*   What are you trying to achieve?
*   How can you achieve it?

- What is stopping you from doing so?
- What are you going to do?

Using these questions can enable you to have a coaching conversation with someone to help them tackle a problem, issue or challenge. Coaches do more than this, though, they also:

- help keep us motivated;
- challenge our thinking;
- give feedback on our performance so that you can improve on it;
- hold you to account.

## The downside of coaching

Undoubtedly, coaching has real long-term benefits in helping to improve safety performance and culture, so it is one of the fundamental new tools that safety professionals must be able to use. However, there are some major downsides that you must be aware of:

- It takes longer to get the result because you have to help someone work out how to do it for themselves, which invariably means that it will take longer than if you just do it or tell them what to do.
- It can come across that you don't know the answer to any question if you always take a coaching style, which can clearly affect your credibility.
- It generally only works when the person is being coached has adequate underpinning knowledge of the subject; if they are not at that level, you cannot really coach them effectively and it will just irritate them.
- It needs to be done at the right time – for example, taking a coaching style when the fire alarm goes off is not the right time to coach people what to do in an emergency; actually, you don't want them to think about anything other than getting out straight away.

## Applying coaching to improve safety performance and culture

The theory of coaching is all very well and good – we have only skimmed the surface of this – but practising in a sterile environment does not really help you develop your skill or indeed give you a feel for how it works in the real world. To do this, then, we will use four examples.

### Safety trainers

In the first example the issue is not about a safety problem but more about a team of trainers the safety professional was asked to manage.

Unfortunately, they lacked confidence to make decisions because of years of working for a very controlling manager. The new safety manager noticed that their team would always ring them at 8.45 in the morning if they were going to cancel a course just to get permission to do so. They decided that this was not the best use of anyone's time and that the right people to make that decision were, in fact, the trainers themselves. At the next team meeting the safety professional told them about this and while they all said they would try in the coming weeks, very little changed. It became apparent that this approach was not going to work and something else was needed.

The next time the trainers called in to get authorisation to cancel the training, the conversation went along the lines of:

*Safety professional (SP)*:   Hi, what can I do for you?

(Coaching question: What do you want to achieve?)

*Trainer*:   I think I need to cancel the training today.
SP:   OK, why is that?

(Coaching question: How can you achieve it?)

*Trainer*:   I've only got three people confirmed to attend and I think I can use my time more effectively today to crack on with my project list.
SP:   Seems reasonable. What's stopping you, then?

(Coaching question: What is stopping you doing that?)

*Trainer*:   Nothing, really. I've spoken to the area manager and they're happy for the three colleagues to go back to work and rebook at a later day.
SP:   Sounds like you've covered all the bases, then. What's the plan?

(Coaching question: What are you going to do?)

*Trainer*:   Cancel the course, I guess.

The safety professional had similar conversations with each of the trainers over the coming weeks. The next time they rang them at 8.45 in the morning, they let the phone go to voicemail. When they called the trainers back later in the morning and asked the reason for their call, their reply more often than not was 'I just wanted to run the cancelling of the course past you but I've done it now'.

In this example all the trainers needed was to have confidence in their own abilities and once they had talked through their justification for the action they were proposing (which the safety professional never disagreed with), over time their behaviour improved. Getting to the end result took a lot longer than simply answering every one of their calls and agreeing with them to cancel the course. However, continuing to do this was untenable and with the free time it helped the safety professional to focus on trying to resolve some of the reasons why the courses needed to be cancelled in the first place.

### Product storage

During a walk round of a site with the manager, a safety professional noticed that there were some pallets of stock stored in front of racking which reduced the amount of space for the site's forklift truck to operate in. Instead of pointing this out to the manager and telling them to get it moved, they took a more coaching style. The conversation went along the lines of:

*Safety professional (SP):*   I see that you've got a few pallets of stock here. How come? Did someone over-order?

*Manager:*   No, we haven't got enough room for them.

*SP:*   Oh, right, but we can't really have the forklift operating in this tight area.

(Coaching question: What do you want to achieve?)

*Manager:*   I can see that. Guess we need to find somewhere else to store them.

*SP:*   Any thoughts?

(Coaching question: How can you achieve it?)

*Manager:*   I need them all stored in the same area as the guys won't thank me if not. I suppose what we could do is remove the third shelf of the racking and that way I could store the full pallets on the second shelf.

*SP:*   Good idea. What would you do with the items already on the third shelf?

(Coaching Question: How can you achieve it?)

*Manager:*   Looking at it, they look like odds and ends, so we can put them anywhere.

*SP:*   Is that the answer, then? Remove the third shelf and move the odds and ends to a better location?

(Coaching question: What is stopping you doing that?)

*Manager:*   Yes, and we'd have to get the racking company in to actually get rid of the shelf.

*SP:*   Great, so when do you think you can get it done?

(Coaching question: What are you going to do?)

*Manager:*   I'll ring them when we get back into the office.

While this was a very simple coaching conversation, it does illustrate how effective it can be. Not only did the manager come up with the answer themselves but the solution is sustainable. Now imagine that the safety professional had taken a more traditional approach to the problem. Would they have achieved the same result? Probably not.

### Yard safety

While working on a safety improvement programme the safety professional went to a location to talk through how they could improve their workplace transport safety arrangements. It would have been very easy for them to walk round the site to observe the traffic routes as well as the people using them and where vehicles were going, and make some recommendations for the site management to implement. However, they were really concerned that if they took this approach, the site management and staff working there would feel as though safety was being done to them and take little ownership of the proposed improvements. Instead, they walked round the site with a forklift truck driver, a lorry driver and the yard manager. They spent time standing and watching what was happening as well as exploring all the options you would when you are carrying out a transport risk assessment.

One of the conversations was in relation to creating a one-way system which would undoubtedly improve pedestrian and vehicle safety. The safety professional believed that there was a way to do this, as did the yard manager as they had been thinking about it since they had attended the organisation's recent safety course. The coaching conversation went along the lines of:

> *Forklift Truck (FLT) Driver*:  I know what you're going to say, you're going to say we need a one-way system.
> *Safety professional (SP)*:  Well, one of the biggest problems we have here is reversing vehicles, a way round that is to put in place a one-way system so that lorries can drive in, unload any returns and move round to the loading bays then out the site without having to reverse.
> *Manager*:  I can see that's a good idea. I think it will help us operationally as well, as we can get in a bit of a mess if the line breaks down and it means sometimes we have lorries in the wrong place. But I guess we are struggling to see how we could do that.
>
> (Coaching question: What do you want to achieve?)
>
> *SP*:  That's fair enough. Shall we have a walk round with that in mind?
>
> (Coaching question: How can you achieve it?)
>
> *Lorry Driver*:  What about moving those crates out of the way and getting rid of the section of fence and put in a gate? There'd be enough room for us to swing out of there.

*Manager*: That would work but what about security? We can't just have anyone wandering into the place.

*Lorry Driver*: Yeah, I didn't think that through!

*SP*: Well, don't give up just yet. Is there something we can do to maintain security?

*FLT Driver*: Place I used to work at had sensors in the floor so when we went up to the door in the warehouse the door lifted as the forklift truck approached and then closed afterwards.

*SP*: I like that. Would it work here?

(Coaching question: What is stopping you doing that?)

*Manager*: I suppose so. So what would happen? Lorries will drive up to the gate, it would open automatically on a pressure sensor in the ground and as they drive through it the gate will close automatically?

*SP*: Yeah, something like that. Certainly worth exploring. How do we do that then?

(Coaching question: What are you going to do?)

*Manager*: I'll have a word with the engineering manager; she is bound to know about this sort of thing.

In this example the working group came up with the solution using their pooled experience and the safety professional simply held the line true to test if there was a way round the security issue. Compare this sort of conversation to a more traditional one about undertaking a risk assessment relating to workplace transport safety. Which one is likely to engage people more?

### Making a training programme

The organisation a safety professional worked for was committed to putting every manager through a two-day safety course. This was a huge investment in terms of both money and time. The safety professional sat in on the first course and asked the delegates what they thought of the course. Unanimously, everyone liked it but when asked how they were going to apply the learning when they returned to their workplace, they were all nervous about it as they knew that the day job would get in the way. At the end of the course they were asked to develop an action plan to help transfer their learning from the course to their workplace. They filled in the happy sheets and off they went.

A few weeks later, the safety professional was in some of the sites whose managers had been on the course and asked what they had done differently since the course, and while they had done some things, the 'bigger ticket items' that would make a real difference had not been implemented.

The safety professional decided that this was a wasted opportunity, so took it upon themselves to call the delegates a few weeks after the course to have a coaching conversation about following through on the plans they had developed. The conversation went as follows:

*Safety professional (SP)*:  Hi, I'm just giving you a call to see if you've managed to put any of the action plan in place. Is now a good time?

*Manager*:  As good as any. I've done some of it, reviewed by risk assessments and stuff.

*SP*:  Excellent, how did you find that?

*Manager*:  It was all right – a bit mundane to be honest, though.

*SP*:  Well done for sticking with it, then! We are planning on revising them in the next few months to make them more interactive. What else was on your plan?

(Coaching question: What do you want to achieve?)

*Manager*:  The thing I really need to do is address my supervisor's ability to lead the team when I'm not here. At the moment they let the standards slip when I'm not about.

*SP*:  What are you thinking of doing there. It sounds a bit tricky.

(Coaching question: How can you achieve it?)

*Manager*:  It is, that's probably why I've put it off. I think I probably need to have them in and explain that I need them to step up and say that if they don't, we will have to move into performance management.

*SP*:  From what you said on the course, I think you are probably right. It won't be easy, but if you can, safety will get so much better at the site and a load of others too, I bet?

*Manager*:  Oh, definitely.

*SP*:  So when are you going to do it?

(Coaching question: What is stopping you doing that?)

*Manager*:  I'm going on holiday in a few weeks, so I will probably leave it until I'm back.

*SP*:  Yeah, but if you leave it until you are back, you are going to knowingly accept that you will have poor standards while you are away.

*Manager*:  You have a point. I might be better off to have the chat before I go.

*SP*:  And that way if things still haven't improved, you are closer to moving to the performance management bit, which means you will get a resolution sooner.

*Manager*:  Yes, you're right.

*SP*:  Do you need any help to do it – from HR, maybe?

(Coaching question: What are you going to do?)

*Manager*:   No, I know what I need to do. My supervisor is back in tomorrow, I will have a chat with them then.

*SP*:   Great. So shall I give you a call in a few days to see how it went?

*Manager*:   That will be fine.

In this example the safety professional had a really good positive coaching conversation, as not only did they get the manager to come up with the course of action they were going to take, they worked out if there were any blockers, they challenged their thinking, gave them some feedback, motivated them and held them to account.

In truth, if the safety professional had not made that call, the chances are that the manager would have continued to accept the poor standards the supervisor allowed. The pressures of the day job got in the way, but the follow-up call that prompted action and the fact the safety professional was going to call back in a day or two to see what had happened spurred the manager on to actually do something. Arguably, the manager's manager should have stepped in to resolve the situation as well. However, they too were busy and sometimes a coach needs to be independent from the person being coached, otherwise it can feel as though you are being performance managed yourself, which is not what coaching is about.

### Coaching a senior leader

As part of a programme to improve the quality of safety conversations that senior business leaders in the organisation have with front-line colleagues, a safety professional was asked to coach a business leader to improve their style. Having spent some time beforehand with the business leader, the safety professional joined them on a planned safety tour. To start with, the safety professional just watched and listened to what was happening.

They arrived at the depot in time to see a group of workers loading up their van. The business leader climbed into the vehicle and started to have a look at the kit in there. They came across a number of loose tools which were stored in an unsafe way.

The safety professional observed the following scene:

*Business Leader (BL)*:   Is this your van, lads?

*Charge Hand (CH)*:   Yeah, what's up?

*BL*:   I'm doing a safety tour and I've just found this (pointing to the poorly stowed hand tools).

*CH*:   Errh?

*BL*:   If you have to stop suddenly, they are going to be everywhere and anyway, what are they doing in the cab area? They should be in a tool box in the cargo bit.

*Depot Manager*:   Yes, good point. We will sort that out.

As the business leader left the vehicle, the charge hand and their two colleagues muttered a few things to each other, loaded the rest of their equipment and left the depot.

The rest of the tour carried on in that vein. At lunchtime the business leader and safety professional went for lunch together to discuss how things had gone in the morning. The conversation went as follows:

*Safety Professional (SP)*:   How did you think that went?

BL:   Really well, I think. I met most of the team there, spotted a few hazards, said well done for having a clean depot – yeah, not too bad at all.

SP:   For me, I wondered if there were a few places where we could have got a bit more from them if we went about it differently.

BL:   Really? I've been doing this sort of thing for years, but go on, tell me how you'd do it.

SP:   OK, what is it you want to achieve with your safety tours?

(Coaching question: What do you want to achieve?)

BL:   I want to make sure people are working safely and that I show them how important safety is to me.

SP:   So how can you do that without doing what you did?

(Coaching question: How could you achieve it?)

BL:   I don't know; I thought you were going to tell me that.

SP:   Fair point, I find that with this sort of thing it is good to try not to think that you have to find something wrong. I often ask them to explain what they are doing and why things are the way they are. I tend to get more out of it that way. What do you think?

(Coaching question: What is stopping you?)

BL:   OK, I suppose.

SP:   So why don't you try it on this afternoon's visit and see how we go?

(Coaching question: What are you going to do?)

BL:   I don't mind giving it a go, but I might want you to lead one part of the conversation so I can see it in action.

SP:   Sure.

This coaching conversation is typical of many you will have. To move into a developmental role you need to watch and listen to what is happening, then offer the feedback. Feedback should be done in the moment where possible and use concrete examples. According to research such informal, accurate feedback can improve performance by 39 per cent (1).

The conversation can then turn into a coaching one, but there could come a point where you need to change your style to one of being a mentor, where you impart some of your experiences and knowledge to the other person for them to think about how they could use it to help themselves. As soon as you have done this, you need to move back into being the coach, otherwise you run the risk of reverting to being a teller.

## Conclusion

The importance of being able to use coaching techniques to improve safety cannot be overstated. It is hugely important. When the circumstances are right, adopting a coaching style helps drive engagement, enables you to learn more, and encourages improvements to be sustainable and longer lasting than other styles.

The best coaching conversations are those where the person being coached does not know it, as the conversation appears normal and seamless. Coaching is a style that is really versatile and once you can guide people through such a conversation, the subject becomes almost irrelevant as you do not need to be expert in it, just to be really good at coaching.

## Reference

1  Corporate Executive Board, *Managing for High Performance and Retention*, Arlington, VA (2005).

# Part 3

# Overcoming inaction

Engaging the unengaged is probably the hardest challenge that safety professionals face and there is no one sure fire way of doing it. If there was, whoever thought of it would be a very rich person. The age-old problem of overcoming inaction is often a question asked in job interviews for health and safety people and can be a source of much frustration for safety professionals the world over.

Altering the way you work using the techniques we have already discussed will help you win over some people and make you a more effective safety professional. In the following chapters we will explore a number of ways that, more often than not experience shows, help overcome inaction at all levels of an organisation. When you add these things together, not only can you overcome inaction, but you can become a role model for others in the organisation of how to get things done; before we can get to this point, though, it is important to set the scene.

## The traditional approach to getting action

As safety professionals, we are taught that there are three reasons why health and safety should be taken seriously because there are:

- legal obligations placed on employers and individuals;
- moral considerations;
- financial benefits in managing health and safety effectively.

While these are definitely valid reasons why people should manage health and safety, there are, however, a number of downfalls with these arguments, particularly if you are addressing people who really do not want to do anything.

### Legal argument

Laws are put in place to make people do something and are enforceable by various law enforcement agencies. Take as an example the use of hand-held mobile phones while driving. It has been an offence to commit this unsafe act in the UK for some time, yet many people still do it. The reason they do it might be because they perceive the chances of being caught are low and even if they are caught they may think that the penalties are insufficient to make them change their behaviour. Unscrupulous employers may think along similar lines when it comes to health and safety in the workplace.

### Moral considerations

While nobody wants to be the one who has to go and knock on the door to explain that someone's loved one has had an accident at work and will not be coming home again, it is questionable how long that mind set will last unless you have been in the unfortunate position of having to do so. Of course, for this mindset to 'stick', it means that somewhere along the line people have to suffer, which is not what we are about.

### Financial benefits

Making justifications for action based on financial benefits for the organisation is a far better solution to the problem, but even then it is all too easy to be somewhat naive when making the commercial argument; as we have already discussed in Chapter 8, simply linking the need for action to reduce sick pay may not result in tangible cost benefits.

To really get attention using this argument, the safety professional needs to be commercially aware in making sound business cases that have positive spins not only for health and safety but for the wider business as well, to the point where improved safety is almost the by-product of reducing the cost base, increased operational efficiency, customer service and sales. However, does it matter as long as the job gets done?

## A different approach

In 1968 Dick Fosbury changed the high jump forever. He managed to jump higher than the other competitors in the 1968 Olympic Games by using a different method of jumping. Instead of the 'straddle jump' that his peers were using, he developed a 'flop' that became known across the world as the 'Fosbury Flop' and is still used today. This example shows that thinking outside of the box really does work. However, while the theory works, it is sometimes hard to do, particularly when you have been brought up to believe that there is only one way to do something.

In overcoming inaction we need to turn things on their head. We should ask why people will not do something rather than telling them why they should do it. Essentially, we need to remove their arguments for not doing something.

Less experienced safety professionals seem to make the common mistake of complicating everything, whether that is having a convoluted system in place or by having reams and reams of paperwork for people to complete. The message is clear: keep things simple, the more complicated things are the less likely people are to do it.

It also becomes embarrassing when safety professionals mention the law when trying to convince people to do something; what the safety professional is suggesting should make sense to the organisation without reference to legislation, if it does not, then there is generally another way to manage the issue.

In reality, though, most people will want to have a 'go' at getting health and safety right. Nobody in their right mind wants to be responsible for hurting someone else. As such, they tend to have a 'go' at it, but because they see it as a great big thing, they chip away at one corner of it and as they may lack the skills and tools to deal with it effectively, they do not get very far and give up. Put like that, who can blame them? The safety professional's job is to break the thing down into bite-sized chunks and drip-feed people.

The idea of avoiding over-complication and mentioning the law as well as providing bite-sized chunks are fundamental, sensible approaches. However, there are other bigger levers we can pull to overcome inaction and make sustainable changes to safety culture and performance. The good news is that they are within our own gift to do so, and while there might only be three chapters in this part of the book explaining this, experience continuously shows that there are incredibly powerful ways to overcome inaction towards safety.

# Chapter 16

# Getting smarter

Many safety professionals spend their whole careers trying to find the 'silver bullet' to overcome inaction towards safety and get extremely frustrated when they fail to do so. Experience would suggest that there is no silver bullet. There is though a collection of bullets that, when used in the right way, with the right people, in the right situation have the desired effect. The skill then becomes to know which tool to use when.

In this chapter we will explore a number of tools that form the basis of this with the following two chapters looking at more 'advanced' techniques and while they are written from the perspective of overcoming inaction towards health and safety, experience shows they work for almost any subject.

## Understanding the audience

The first and absolutely critical step to overcoming inaction is to know your audience and the place they are in. Even in the most unengaged organisation there will always be some people who are interested in health and safety. The thing to do is to identify them as well as the other key players. This can be achieved by performing an 'engagement-influence' exercise. The exercise is a standard, commonly used management model and places key players appropriately on a graph (see Figure 16.1) in order to work out how much effort you should place on them based on their engagement with safety and influence in the organisation.

The goal state with this model is to have everybody, irrespective of level of influence, showing high levels of engagement. Once people have been placed on the model it is important to use the results to help achieve the goal, for example:

- High influence–high engagement and low influence–high engagement. People in these categories are where you want them to be. The key action to undertake with these people is to maintain their interest and momentum.

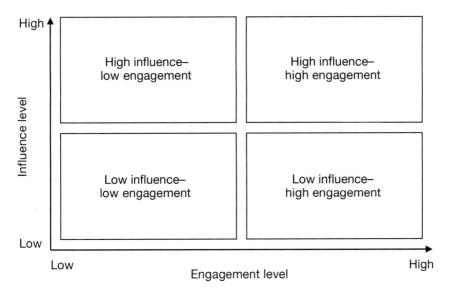

*Figure 16.1* Influence–engagement model

- High influence–low engagement.
  Your attention should be focused on people in this group. These people are the ones with the real influence in the organisation, yet they are not switched on to what you are trying to do. This is not good.
- Low influence–low engagement.
  Although it is not good that this group of people have lower levels of engagement with what you are trying to achieve, for obvious reasons getting them on your side is a lower priority than those with more influence.

This technique must come with a warning. Be very careful where you keep your completed influence–engagement model workings, because should others see them and the contents get out, it would be like dynamite and will probably ruin your relationship with some people as well as almost guarantee that you will not make sound ones with others. The best advice is either to shred them or have them mentally in your head.

Having identified who you need to work on, it is important to try to understand the people individually by finding out about their personalities and motivations, as these will give clues as to how to approach them and even find common ground. Often common ground (inside or outside work) is a good ice breaker and helps people to warm to each other from the beginning.

Moving beyond these basics, though, is important, as you need to find out what are their drivers – what makes them get out of bed in the morning. Is it sales, reducing cost, maximising productivity; what is their 'thing'? Armed with this information you can start to find solutions to safety problems in the organisation linked to their particular interest to start the process of engagement.

This is probably best illustrated with an example. A safety professional wanted to develop a programme to help delivery drivers improve their situational awareness when they deliver to a site, in order to help tackle an increasing accident trend in the organisation. They knew from their influence–engagement model that the operations director was highly influential but not as engaged as they would like him to be. Instead of selling the virtues of the programme from a safety perspective, they approached it differently. The operations director was heavily focused on customer service, so the safety professional used this as a lever to get the result they wanted. An accident had occurred recently and the customer had emailed the organisation expressing their disappointment that the incident happened and outlined the knock-on consequences to their organisation as a result. Making the link between improving the driver's situational awareness and customer service became much easier and more tangible. Furthermore, the safety professional knew that the operations director loved flight simulators, so they engineered a way for the programme to be delivered by the use of a 3D simulator. While the conversation was still difficult, the safety professional managed to get the programme moving, whereas if they had approach it from a more traditional standpoint, they probably would not have succeeded.

## One of the team

As we have discussed elsewhere, in some organisations the safety professional and function will be part of the 'operational line' or sit independently, and there is nothing wrong with either. If you are part of the team, or seen to be part of the team that have the safety problem or issue and you work together to resolve it, you are far more likely to come against fewer people who are resistant or whose resistance is lessened.

This is particularly well illustrated in the following example. An organisation underwent a structural change where their safety professionals reported into the operational line rather than to a central team. The senior safety professional for each part of the business sat on their leadership team as a peer to those who led the other functions. Over time, they were seen and accepted as 'one of our own' to the point where, while there were healthy, challenging conversations about safety, they always came to the right solution in the end. Whereas when other central functions came to ask for support or present a new way of working they received far more challenges and quite often were sent away with 'more work to do before it will be taken forward'.

If you are already part of the team, from a reporting perspective, things are easier, but if you are not, then you need to see part of their team. Start with the leader, the person who is the 'boss' in that team. Spend time with them using the approaches we have already talked about in the previous section. Then try to really understand the organisation, what are the pressures the others in the team are facing, what are their priorities, and begin to link ways to improve safety to ways they can solve their other problems or priorities – again, as we have discussed. To get to the end result, it will take a while, especially if there are tricky personalities in there.

Another, more risky approach, is to take one for the team. Again, this is explained better through an example. The team a safety professional found themselves trying to 'get their way into' was, at best, described as dysfunctional and did not like to let anyone in. There was a change programme on the horizon that the team did not agree with, for very valid reasons, but it was being driven by the safety professional's boss who sat centrally in the organisation. Despite their best efforts, the safety professional could not get the team they supported to change their view (a view that they shared personally) and during a business review took a real dressing down from their manager in front of the rest of the team for the stance they were taking. Clearly, the behaviours of the safety professional's boss were wrong. However, as soon as the meeting was over and they walked out of the meeting room, the team's attitude towards the safety professional changed. The relationship was never one of 'best friends', more one of respect and appreciation, which at times is all you need.

## The right language

As discussed in Chapter 1, the safety profession does tend to suffer from a stigma of being very 'anal' and have a clipboard and cagoule approach. It is important to break this mould and to talk in the language that is appropriate to the people you are addressing.

Obvious examples include when talking to directors and senior managers they tend to converse in 'business speak' – about helping to widen the margin, reducing the cost base, advancing the people agenda, driving top line sales – while talking to front-line managers and employees it is important to communicate without fancy words.

As silly as it sounds, you would be surprised how using the right language can help you switch people on to what you are talking about; sometimes it is about the other person positioning you differently and having seen it happen on many occasions, the right language is very powerful.

## Being savvy

Understanding how organisations work is important. For example, going along to an operations meeting and expecting to get all members of the meeting (including their boss) to agree to something that you present for the first time is at best optimistic, at worst naive.

Playing dot-to-dot is vital and while going round talking through and selling the idea to each member of the meeting individually might take time, it will give you the opportunity to build into the proposal countermeasures to their concerns before they air them in front of the colleagues at the meeting. Not only does your credibility remain intact, but your relationship with each of these people grows exponentially.

It is also time well spent to think about what the counter arguments are likely to be when you meet the people you want to influence. If you can think of them beforehand and have your answers ready, this will help no end. As an example of this, a safety professional had recently joined a new organisation and was not comfortable with the risk assessments they had in place and wanted to improve things. The challenge they had was that they had to convinced five operations directors and the group operations director that this was the right thing to do. Coupled with that, the place where the decision would be made was their monthly operations leadership team meeting, which was a meeting that other support functions had been asked to leave because they did not like what they had said.

Knowing that going in cold was never going to work and only serve to prevent them building any credibility in the long term and that the perception was that the assessments must be acceptable because the organisation had never been in trouble as a result of them before, the safety professional decided on a different course of action.

They arranged a day out with one of the five operations directors. As part of the day they asked them for some help as they were struggling to understand how the current risk assessments worked as the approach was quite different from what they had seen before and they wanted an explanation. The operations director did their best but they too got confused and at this point the safety professional showed them what an alternative might look like. The operations director made some suggestions that they would like to see with the approach and they went on with the rest of the day.

The safety professional incorporated the first operations director's tweaks and replayed the same approach with the other operations directors, slightly tweaking things each time. A week before the operations leadership team meeting, they sat down with the group operations director and explained why the assessments needed to change, dropping the line 'even the ops directors aren't clear on how they should use them, so what chance do our front line managers have?' and showed the alternative. With the group operations director's blessing, they presented the proposal at the meeting, and they were amazed when the response from the room was 'I thought we agreed this before; when do you think you'll have it done?' (1).

## Work as a team

A common throw-back from people who do not want to do things is that the safety professional does not know how the operation works in the real world. Therefore, if you go in telling them what to do, the barriers will come up instantly. While this may not be the case so much with safety professionals who have risen through the ranks or who have moved sideways into the role, it will certainly be a consideration for the growing number of people who are entering the profession as a first career choice.

From experience, one of the best ways to overcome this challenge is to be honest and admit that they are the experts in what they do (whether that is quarrying, warehousing, etc.) and you are the specialist in health and safety. The two parties should work together to make the most of each other's expertise. At the end of the day, most of what they want to do operationally will be fine from a health and safety point of view. It is normally only a small element that will require a little more work to make it acceptable to both parties.

One of the best ways to gain respect, build your credibility and help with this approach is to regularly put yourself in a front-line colleague's shoes. Take a day or two out of your calendar and do the job they do. Without doubt you will see the effect of your work (good, bad or indifferent as it might be) and you will really know what happens.

## Dealing with the 'not my job' argument

At some point someone will declare that health and safety is not their job and they do not need to do it because that is why the organisation has a safety professional. While this view is clearly wrong, arguing the point based on the duties of some statutory instrument might not get you very far, particularly given what has been discussed previously in this chapter.

The best solution is to put it into context, as an old boss did very well with an operations manager who was somewhat against doing risk assessments, the conversation went along the lines of:

> *Safety Professional (SP)*: 'When you leave your company car do you lock it?'
> *Operations Manager (OM)*: 'Yes, of course I do.'
> *SP*: 'You don't get a security guard to do it for you because really that is a security issue.'
> *OM*: 'Don't be silly!'
> *SP*: 'Well, the same thing applies with health and safety!'

## Conclusion

It is accepted that on first reading some of the thinking and examples might seem a bit odd or even a step too far for those with a more traditionalist approach. However, if you think that, ask yourself honestly whether you have a problem overcoming inaction towards safety in your organisation. If the answer is yes, then you either have to change those people who are resistant, or you need to change your approach to engage them. While it is a big ask for you to change your approach, it is easier and far more likely to happen than to replace all those people in the organisation who are reluctant, and arguably in most organisations it tends to need to be an extreme situation to remove someone from their role just because they are resistant to health and safety.

The simple tools we have explored in this chapter really do work. You need to take your particular problem or issue and think about how you can tackle it using these principles.

## Reference

1   www.shponline.co.uk/on-the-right-track-dealing-with-sceptical-management/ (accessed 21 November 2015).

# Chapter 17

# The law of diffusion on innovation

Have you ever wondered how firms get people to buy their goods or services to such a point where they become hugely successful businesses and market or even global leaders? There are lots of things that need to come together to make this happen, and a fundamental aspect is the marketing campaign that supports it. Don't worry, this book is about health and safety, but the point is that we can learn a lot from sales and marketing professionals, about how they achieve mass-market penetration of their products and apply those same principles, surprisingly easily, to overcome inaction towards safety (and just about anything else too).

In this chapter we will look at one very powerful way you can 'sell' based on a theory known as the law of diffusion of innovation.

## Diffusion of innovation

The law of diffusion of innovation is not new; it was first articulated by an American Professor of Communication Studies in the early 1960s (1) and has been retold several times since (2). The approach breaks the target population down into five subgroups:

- **Innovators**   The people who come up with the idea.
- **Early adopters**   The people who just get 'it' – in other words, they are people who will queue up for hours to be one of the first to get their hands on the latest computer game release or electronic gadget.
- **Early majority**   The ones who want to see what all the fuss is about.
- **Late majority**   The people who follow suit because everyone else is doing it or have it. They are asking themselves what they are missing out on.
- **Laggers**   People who get the product or service because no alternative is available any longer.

This is commonly shown in a graph, like the one in Figure 17.1, which also shows the percentage of people who account for each subgroup in any

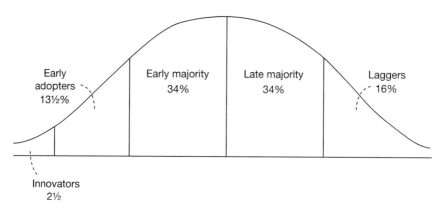

*Figure 17.1* A commonly used graphical representation of the law of diffusion innovation

given population. Typically, according to this approach, there is the need to get about 15–18 per cent of the given population interested, and once you do this you achieve the 'tipping point' or 'critical mass' where things just snowball as the momentum keeps building (2).

## Applying it to safety

The obvious thing to explore now is to think how this approach can be applied to overcome inaction towards safety. The first thing to realise – and this is the clever bit – that once you get past the early adopters to everyone else it will not be an idea driven from the safety professional. It has come from someone like me, it is something my peers have done and they look like me, they sound like me, they do the same job as me and it might just work here too.

Let us explore how you can put this into practice through five examples.

### A new way to reduce lifting

Following on from a manual handling training programme, a location decided to raise the container they receive their deliveries on to waist height by resting it on three good quality and lashed together wooden pallets. This dramatically reduced the amount of lifting and bending people had to do.

The organisation's safety professional happened upon this one day and was so impressed, added it to a list of success stories that listed all the different things people had done as a result of the programme. They circulated the stories around the network and on their other site visits were pleased to see that others were taking the raised container idea forward too. Some people still did not get it, but more and more people did.

Six months later they were at the opposite end of the country from where the idea had come from and saw it in action. When they asked about it, the front-line colleague said, 'Oh yes, it was an idea from such and such'. This simple example shows that while in most cases the safety professional will be the innovator as you will be the one that comes up with the safety idea, process change and improvement programme, sometimes all you need to do to be an innovator and get the snowball rolling is to recognise something that is good and promote it.

### Operational middle managers breakfast chats

An organisation undertook an attitude survey of their workforce asking them for their opinions on how they manage safety. One of the key findings was that the organisation's middle managers were not visible enough on the front line and they do not listen to front-line colleague's safety concerns.

As is often the way, a working group was set up to tackle the matters arising and, despite other suggestions to address this particular piece of feedback, the group made an objective for every middle operational manager to go and spend time at the work location of front-line colleagues. However, this just turned into them basing themselves at the location for the day rather than engaging with those working there. The safety professional saw what was happening and went to one of the middle managers who they previously identified as being quite engaged with safety and had a good level of influence with their peers (see Chapter 16 for how to do this) and who, as it turned out, did not like the approach.

They discussed one of the proposals that was originally discounted by the group, which was to have breakfast chats with front-line colleagues where the manager would take some bacon rolls and drinks and host a breakfast session in the mess room where they would talk to the group of colleagues and answer any questions they might have. On further investigation it seemed that the working group had discredited this idea because they were worried they might get into a difficult discussion with colleagues that they would rather not, or not know the answer to a question they were asked.

The safety professional and the early adopter explored this further and after some coaching they arranged the first breakfast chat. The manager had a list of things they were going to talk about and some questions to get their team talking to them. The session was reasonable and the next time they did it with another of their teams it improved and so it kept going.

The manager found out a number of other nuggets of information from these sessions other than the safety issues their people faced, and so started to get improvements with not only safety but also operational efficiency as they resolved the challenges. In talking to one of their peers they mentioned this to them and they wanted to find out more. The manager explained what they did and invited them to attend a session to see what goes on. Once they had, they went and set up their own breakfast chats, except they did it at lunchtime with pizza.

Momentum started to build with eight of the operational middle management team doing them; at a six-monthly review of the actions arising from the survey they explained what they were doing and now the two managers who were on the working group who said it would not work had to adopt the approach as they were not getting anywhere with the route they had taken.

This example shows two key things about the law of diffusion of innovation applied to safety:

- It can take quite some effort to build up to the tipping point but once you get there it moves forward on its own and all you need to do is to help steer it.
- It can also take a significant amount of time to get to a point where the laggers adopt things. For this reason, you need to make sure you use the approach wisely and think about things as a marathon and not a sprint.

### Change network

A safety professional had just started with a new organisation and soon saw that there was a great deal of apathy towards safety at the site from front-line colleagues. From spending time talking to them, it turned out this was because they feel they do not have a say in how things are done. As it happened, the site already had a safety committee but it did not really do a great deal other than to fulfil legal requirements. To the safety professional this meant that they already had a group of early adopters to safety – they just needed to find a way to get their attention.

In their first proper sit down with the site manager, the safety professional explained what they had seen and how they wanted to tackle it. The safety professional had already picked up that when the forklift truck contract was being retendered and the existing trucks would be replaced, there would be a great deal of disquiet from the users. With all this in mind, they agreed that members of the safety committee could be the ones to trial the new trucks and their feedback would be included in the decision-making process.

When this was explained at the next committee meeting there was some surprise in the room and they were more than a little sceptical that it might not happen, but they were also energised by the idea. The site manager and safety professional followed through with their promise and the committee members, who operated a lift truck as part of their job, got to try out the different trucks they could end up with. They fed back into the group making the decision and were pleased when they were listened to.

When the new trucks arrived, the site management team were expecting lots of push back from their teams and while they did get some, it was not as much as they were expecting. The safety professional asked the committee members why this was the case and was pleased to learn that their friends had been asking what they were doing on the new trucks and what they thought of them.

To help quell the noise further, the safety professional arranged for members of the committee who took part in the trial to go along to different team briefings and talk about the trucks, why they liked them and how they found them to operate. This peer-to-peer interaction, coupled with some other positive voices in the teams, helped move the site's view of the new trucks towards the mass-market penetration end of the scale.

This example shows three more key things about using the law of diffusion of innovation to overcome inaction towards safety:

- The approach applies to front-line colleagues just as much as it does to managers and leaders.
- Often you have ready-made early adapters you just need to explore all the different options you have. Admittedly, this is a little harder if you have a multi-site operation, but you generally do hear about those people who 'get it' and want to be involved.
- Moving from that second 13.5 per cent to the next 34 per cent in the early majority sector can just be as simple as a chat at tea-break time, provided it is a peer-to-peer conversation.

### Getting traction

An organisation was split into five divisions and each had a safety professional. For some time the organisation had set the various area managers within each division personal objectives related to their area's safety performance. However, this had not driven safety forward as had been hoped, so they shied away from setting safety objectives as a result. This also meant that despite their best efforts, the level of engagement with safety by the area manager team had fallen.

One of the safety professionals decided that as part of their division's safety plan, each of the area managers should lead on an element (with the necessary safety support) and they should be judged on the output of the improvement activity. The divisional director agreed to this, so the next year they used this approach.

Midway through the year they were having lunch with one of their peers from another division and explained what they were doing. They were still struggling to get traction and wanted to see how things developed with the approach that the safety professional was using. Needless to say, the amount of traction they achieved was much higher than the existing method.

At the end of the year both of them took the same approach in their divisions – i.e. area managers having output objectives related to safety improvement programmes. By the end of the next year, the other three divisions adopted that approach, but in one case this was not because the safety professional thought it was a good idea, but because the divisional director did. Again, this example shows us several things about the law of diffusion of innovation with respect to overcoming inaction towards safety:

- sometimes it is another safety professional that can be the lagger and will only move when they are forced to by their line manager – while this is not ideal sometimes it is necessary;
- the approach is about influencing people, it applies to getting front-line colleagues to take ownership or accept something as much as it does to getting safety professionals to move.

### Changing people's outlook

A manufacturing organisation was getting into backlog with its maintenance programme as they could not get access to the machinery when they needed it because of production pressures. There were also a growing number of minor accidents relating to smaller non-invasive maintenance work that was not being done because the maintenance team were losing on average 30 minutes per job while waiting for the machine to be handed over to them.

Picking up on this, the safety professional asked the maintenance safety representative when they were waiting for the machines to be handed over, if they could do some of the non-invasive work. When the safety representative realised the issues it was causing they agreed to do it, but said they might struggle to get their colleagues to do it. The safety professional got them the list of non-invasive maintenance backlog and the next time they were waiting for access, the safety representative started to share the jobs to the others in the team. A few of the group started on the outstanding work and after it became apparent that the line was not going to be handed over for some time, the rest joined in.

Over time, they became more accepting to this and other shifts followed suit. It was also interesting that the other teams, mainly production and dispatch, changed their opinion of the maintenance workers from one of being lazy (as they always saw them hanging around doing nothing) to one of 'we are all in it together'.

This is another example that shows us some points to remember about the law of diffusion of innovation:

- Knowing who your early adopters are is critical, and safety representatives or similar people are, mostly, heaven sent for this sort of thing.

- While in previous examples the effort taken to get to the tipping point can be massive and with a great time delay before you see progress, it can also happen really quickly.
- Getting peer to peer influence should not be underestimated. If the maintenance manager or the safety professional had told the team to do what they did, they would have found themselves fighting to get them to do it, whereas this way they decided to do it themselves.

## Conclusion

Although the law of diffusion of innovation is aimed at how you market products or services, taking its principles to overcome inaction towards safety is easy and experience shows works every time. The trick is to make sure what you want to market is sound, then find the people that will queue up all night to get their hands on the new thing and get them involved. Once you do this they will do the rest for you and it will snowball.

It is this snowball effect that sets this approach apart because quite often with a more traditional approach you do you can get about 20 per cent of your target population brought in and consider it a good result, but in safety if you are ever going to achieve the type of safety culture and performance we all strive for, you need much more than this. The real beauty of this approach is that it costs absolutely nothing except your time and in reality you will probably spend less time using the law of diffusion of innovation than you would using your traditional methods with less success.

## References

1   Rogers, E., *Diffusion of Innovations, 5th edn.* Simon & Schuster (2003).
2   www.ted.com/talks/simon_sinek_how_great_leaders_inspire_action (accessed 14 November 2015).

# Chapter 18

# Transactional analysis

Have you ever found yourself in a situation where you have a really strong argument but you just cannot get the other person to accept it? Sometimes overcoming inaction is all in the way things are communicated, not through reports or presentations but by the conversation – in particular, the language (both verbal and non-verbal) that is used and how it is said.

In this chapter we will explore how we can become more aware of other people's view of the world and how we can alter our approach to make sure we achieve our goal or, just as importantly, learn when to leave it, walk away and have the conversation at a better time. To do this, we can use something called 'transactional analysis', which is a psychological theory that is widely used in all sorts of areas from counselling, training and development, leadership and people management.

## Transactional analysis

Retold countless times and built on since, as well as being applied to different situations, the concept of transactional analysis was first defined in the 1950s by psychiatrist Eric Berne (1). It suggests that when we communicate, we do so via interactions or transactions with each other; in other words, I say something you say something back. How we take in the message and respond depends on which of three ego states we are in at the time: parent, child or adult. Two of these ego states break down again so that you end up with five:

- **Critical parent** If you are in this state you are generally seen as controlling, authoritative or demanding, and come across as being judgemental. Often your eyes are locked on the other person's, leaning forward towards them, frowning or with your arms crossed, as though you are telling them off.
- **Nurturing parent** Typically, in this ego state you provide emotional support for others, empathise with others and offer protection. In this ego state you are seen as sympathetic, you tend to be more doe-eyed,

leaning forward with very open posture, often smiling and nodding as the other person or people are talking, as though you are encouraging them.

- **Adult**   This is the ego state where you are calm and rational, and arguably make the best decisions. You tend to be assertive with strong eye contact, which is generally only broken when you are thinking. Your posture could be either leaning forward or back, but always open as you remain still and alert during the transactions.
- **Rebellious child**   In this ego state you are playful, express your feelings clearly and as the name suggests somewhat rebellious. Your gestures tend to be frequent, energetic and at times over-exaggerated.
- **Adaptive child**   Here you tend to be very accepting of the situation and generally do as you are told. You come across as being submissive, typically avoiding eye contact with the other person during the transactions displaying body language, almost as though you are being told off.

The idea is that if you are aware of your ego state (seen through your outlook towards a particular subject and your behaviour) as well as that of the person you are transacting with and if you are not having a productive conversation, if you move to a different ego state you increase the chances of getting a positive outcome. Or put another way, you can overcome inaction. To explain this, we will look at a typical situation that any parent, or adult for that matter, can probably relate to.

Imagine you ask your teenager to do something. They don't want to and say so. You get frustrated and ask them again, raising your voice slightly. They mirror you by raising their voice but still say no. This continues and escalates into a row that neither person is going to win. Here you are being the 'critical parent' – in other words, saying 'Do as I say' – and the teenager is the 'rebellious child' saying 'No, I won't'. To overcome this impasse something needs to happen; you could move to 'adult' and explain why you want them to do the thing you have asked or you could back down by moving to the nurturing parent state by saying 'OK, don't worry, I will do it'.

There are, of course, different ways you can handle this situation such that either both people win or you get what you want out of it and will explore it later on. This is all very well and good, but how do you use this self-awareness and awareness of others to help you overcome inaction? Transactional analysis suggests that to do this you need to get into the 'adult' state. When both of you are there, you will work through the issue and find a resolution – in other words, overcome inaction. That said, there are other ways too, such as 'massaging' the other person's ego. To do this, we will use a simple example – trying to get someone to implement a new process. Identify which ego state they are in and then massage their ego using the following approaches to get what you want to overcome the inaction. If they are in the following:

- **Critical parent**   Ask for their opinion on the subject – for example, 'How would you go about implementing this new process?'
- **Nurturing parent**   Ask for their help – for example, 'I'm really struggling to get David to implement this new process. Could you have a chat with him for me, please?'
- **Rebellious child**   Offer them something that they want in return – for example, 'If we can agree to introduce this new process, we can then start to look at . . .'.
- **Adaptive child**   Offer them the opportunity to do something to aid their development – for example, 'Implementing this process will be a great chance for you to show what you can do'.

Don't be fooled into thinking that the only productive and useful ego state is the 'adult' one – it is not. Each ego state is really useful if it is used in the right circumstances:

- If one person is in the 'nurturing parent' ego state and the other the 'adaptive child' it is great when care and concern are needed. Take, for example, a conversation between an employee who has just had an accident and their manager. Clearly, their manager being in the 'critical parent' ego state would be unhelpful.
- A transaction between one person who is in the 'controlling parent' ego state and the other who is an 'adaptive child' is the most helpful when you are trying to give instructions. For example, you would not want people at a conference to be in the 'rebellious child' ego state when the fire alarm goes off. You simply want them to follow the instructions of the facilitator so that they can get out as quickly and safely as possible.
- When two or more people in a group are in the 'rebellious child' ego state, they can be quite a handful unless you are at a party and want to have fun or you need to be creative in order to help you solve a problem.

## Applying transactional analysis to overcoming inaction towards safety

We will now look at a number of examples to show how you can apply transaction analysis to overcome inaction towards safety.

### The critical safety professional

A safety professional wanted to introduce a programme of safety inspections undertaken by the organisation's area managers on a six-monthly basis, to help improve their safety assurance that the risks at their sites were being managed effectively and to demonstrate their visible leadership of safety. They had designed the question set as well as a short guidance note that the area managers could refer to when they were completing the inspections. They had agreed with the area manager's director that they had three months to complete the inspections at all of their branches, which was more than enough time.

Unfortunately, on the day when they were explaining the process and why they wanted the area managers to do it, the director was late for the meeting, which meant that the safety professional had to launch the new inspection programme on their own. While they were waiting to start, the conversation between the area managers was all about the new targets and the latest thing to come out of head office, phrases like 'I'm not going to do that', 'What a complete waste of time . . .' and 'I'm not sure they know what they are doing . . .' were being used.

When the meeting started, the safety professional got on to the subject of the new inspection programme and, despite their best efforts, many of the area managers did not want to do it, saying that they had too much to do already and they were now under extra pressure as their sales and productivity targets had just been increased for the rest of the year. The safety professional pointed out to them that they had made the commitment some time ago and that it was in their safety plan as a key leadership activity and therefore they had to do it. The area managers wanted to delegate the inspections entirely to their support managers. However, this was obviously not going to achieve the objectives the safety professional wanted. This circular conversation continued for five minutes, escalating each time. It became clear to the safety professional that they were not going to get things moving unless they changed their tack.

They weighed up quickly whether it was worth massaging the area managers' egos. However, as the managers seemed to be in the 'rebellious child' ego state they decided that there was nothing they could offer them in return for doing the inspections themselves. With this in mind, they decided to go to the 'adult' ego state. They summarised the situation as:

- I can understand you're up against it now that your sales and productivity targets have been increased.
- We also need to make sure we deliver our safety plan for this year as well as driving our safety leadership programme forward.
- I am uncomfortable in letting your support managers undertake all the inspections for you, as clearly it defeats the object of the safety plan.
- We need to find a compromise.

In moving to 'adult' and being assertive in their approach, along with a logical argument that acknowledged the issues of both sides, some of the area managers also changed the state that they were in from 'rebellious child' to 'adult'. One of the managers suggested that they undertake inspections at their branches which they believed to be the highest risk, and they could leave the medium- and low-risk branches to their support managers. This approach was the compromise that was needed so that both the safety professional and the area managers both won, or perhaps more importantly nobody lost. This is a typical exchange that will happen all over the world many times a day in all sorts of organisations. There are some important things that we can learn from it:

- In this example the safety professional was the 'critical parent'. Experience bears out that safety professionals tend to be quite surprised that they come across in this way and hence they are part of the problem in relation to inaction towards safety.
- Moving to the 'adult' state helps diffuse the situation and, in doing so, forces the other person either to mirror you and go to 'adult', or continue as is but then it is obvious to all that their approach to the situation is unreasonable.
- Look out for tell-tales in the exchange that suggest which state both you and the others are in. When things escalate quickly, learn to change tack. In hindsight, it might have been better for the safety professional not to get into an exchange in the first place, particularly as the group had given off enough signals before the meeting started that they were in the 'rebellious child' state.

### Someone working unsafely on site

A safety professional was walking through a site when they saw someone working unsafely. They went over and asked them to stop what they were doing. They then got into a conversation with them about what the person was doing:

> *Safety professional (SP):* You do realise, don't you, that what you were doing then was really unsafe; you could have quite easily got your hand trapped in the machine.
>
> *Worker:* No, I didn't, sorry.
>
> *SP:* What do you mean you didn't? You've been trained, haven't you?
>
> *Worker:* No, I haven't. I only started here two weeks ago. I'm still learning.
>
> *SP:* That's not good. I can't believe we've let you come and do this job without any training. I'm sorry about that. I will pick it up with your boss. In the meantime, isolate the machine before you start any work on it.
>
> *Worker:* OK.

Having spoken to the worker's line manager, the safety professional walks back to where they were going originally and sees that the worker has not isolated the machine and is again working unsafely. This time the conversation goes along the lines of:

> *SP*: Oi, stop. I've just told you to isolate the machine before you work on it.
> *Worker*: Oh, yes, sorry. I forgot.
> *SP*: Right, well, come on, you've got to pull your socks up.

Again, this is a common type of conversation that safety professionals, and indeed many others, can get into on a daily basis and, as before, there are a number of important things we can learn from this example:

- In both parts of the exchange the safety professional was in 'nurturing parent' and the worker 'adaptive child'. In the first set of transactions this is probably the right approach, however in the second one it is not. After all, if you were the worker faced with that, how likely are you to improve? Arguably, not very likely as you have not changed your behaviour from a few minutes before.
- In these situations you have to ramp things up. There was nothing wrong with the first part of the conversation, save for the fact that the safety professional should have taken the worker off the job until they had received the right training, explaining why they were doing it. That would have been the safety professional moving into 'adult', preventing the subsequent unsafe action.
- Overcoming inaction towards safety is not just about getting managers or organisations to do something. It is also about getting those who will be injured or suffer ill health if they work unsafely to change their behaviour. The tack you take will determine if they take on board what you say and whether they change their behaviour as a result or whether you come across as being 'clipboard and cagoule'.

### Winning over the board

A safety professional was taking a paper to a board with a view to getting them to sign off a major change improvement programme. Their argument was sound but they had heard that their predecessor had always struggled, often failing, to engage the board to implement change, so was expecting a rough ride.

In planning their session they prepared a short presentation that complemented the paper they had already submitted and outlined the facts of the matter:

- we need to radically alter how we manage risks associated with people working at height;
- this is because we have already had a very close call where one person fell through a skylight;
- we are seeing more regulatory focus on us as a result.

They then explained what the organisation needed to do and how much it would cost. While there was some discussion about how much money the organisation needed to set aside to tackle the issue (which is always going to be the case), the safety professional was surprised just how well things had gone, particularly when the board agreed the paper.

Afterwards, they sought out their director and said this to them. In the conversation that followed it came to light that the previous safety professional always came across in the board meeting as submissive and asked for permission to do things, which gave them the impression they were not sure of what they were asking for and why. This meant that the board were seen as blockers towards such things, but the truth was that they were just frustrated, so individual functions went off to do their own thing.

This time around the safety professional was 'adult', not 'adaptive child'. They outlined the situation and made a solid argument as to what needed to do be done and why. They were not arrogant, just stated facts respectfully and responded in a balanced way when challenged.

What this example shows is that:

- When you are in the 'adult' state it is important to be careful that you do not come across as arrogant or bullish as this will make the others you are transacting with alter their state and you will end up moving towards an unproductive conversation.
- Sometimes people who you think are blockers or not acting positively towards safety, are misunderstood and you should treat every interaction as a new one and try to forget any history.

## Conclusion

Inaction towards safety comes in many forms. Experience shows that sometimes the attitude of the safety professional and, the way they react to the person they are interacting with, can create a situation where inaction develops. Transactional analysis helps to overcome this.

We all have different preferences for our ego states depending on how we feel towards the subject and the person or people talking with us. When coming up against inaction, it is worth thinking which state you are in and which state the other person is in and asking whether they are opposing or unhelpful in getting things done. If they are, someone needs to change.

It is worth saying, though, that starting to view conversations in terms of transactions initially can be challenging, typically because we have not been brought up to see things in this way. Yet when you start to look out for the tell-tale signs of the different ego states the other people are in, you will be surprised by the results you get when you alter your state either to get to 'adult' or, if appropriate, massaging their egos.

## Reference

1   www.ericberne.com/ (accessed 14 November 2015).

# Part 4

# Engagement tools

So far we have talked about you, the safety professional, and how your approach influences safety. We have explored various tried and tested general management and leadership techniques which, when applied to the role of the safety professional, help increase your ability to become really effective. We have also talked about how you can overcome inaction towards safety and that dealing with this is crucial to success. However, we have not yet talked about the 'game changers' for an organisation's safety performance and culture. Anyone can put together a basic safety improvement programme or design a policy and procedure. Using the things we have talked about so far will increase the chances that they will be much more effective and likely to bring about change which will be seen in an organisation's safety performance and culture. However, the real game changer is about getting everyone in the organisation engaged with safety. If you fail to do this, all you will do is create the illusion of change. Experience shows that lasting change is driven by:

- focusing on the right things (having the right strategy);
- doing things in the right way (using the rights tools and achieving buy in);
- finding practical things for all levels of the organisation to do in order to help you deliver the strategy.

In the next three chapters we will explore some of the ways that experience shows really does drive engagement with safety to help deliver lasting changes in safety performance and culture. The good news is that in the main they are really simple and do not require huge amounts of money. The tools have been loosely grouped into those which appeal to business leaders and senior managers, middle and junior managers, and then front-line workers and managers. That said, when you read through them you will see how they are interlinked, which is fine, but remember that safety seems to be one area of organisational life where people work one level below where they 'sit'. This means that it is important to refocus people who fall into this trap;

most of the time they do so because they are not sure what they should be doing instead, so the following chapters help provide some alternative activities.

There are a few things to bear in mind with the tools we are going to explore:

- They are not an exhaustive list. There will be other things that people and organisations do to affect real, sustainable improvements in their culture and performance.
- You should not think that all you need to do is implement all of them and you will have an amazing culture and performance record. It is about you using your knowledge of your organisation and deciding which ones are right for it at the time.
- Don't forget that you absolutely must combine the tools with the things we have already discussed. Remember, there is no point having the best programme if you cannot, for example, prove the business case for funding or you cannot overcome someone's inaction towards it.

The key thing experience shows in creating sustainable change in organisational safety culture and performance is to focus on a handful of things that will make the biggest difference and then be relentless in driving them home through positive engagement.

# Chapter 19

# Business leaders and senior managers

It should be immediately obvious why it is important to have business leaders and senior managers engaged with safety and the link this has to improving an organisation's safety performance and culture. Traditionally, from a safety perspective, such members of an organisation tend to be little more than figureheads – for example, a signature on a policy statement or a picture in a handbook, and while this is a hugely important part of their job, they can easily do so much more with very little effort. Most of the time this is not because they don't want to, it is because they are sometimes unsure what they could do to help.

As we saw in Chapter 2, the climate towards safety is changing and recent changes in legislation, coupled with the drive to be seen to be doing the right things, means that more and more business leaders and senior managers want to do something in this area. However, if they do not get sufficient direction about what this should be, their efforts could be misguided or at least not as effective as they could be.

In this chapter we will explore some simple yet 'game-changing' things that business leaders and senior managers can do to help improve their organisation's safety performance and culture (building on the senior management audit that we explored in Chapter 7) and, crucially, they can be seen to be doing them.

## The role of business leaders and senior managers

There are some unique characteristics about a business leader's role and that of a senior manager in an organisation, and understanding this is the first part of effectively engaging them. The following is taken from the Institute of Directors fact sheet (1) on the subject and outlines that they:

- provide leadership for the organisation;
- determine the future of the organisation, its ethics as well as protecting its assets and reputation;

- consider the organisation's stakeholders in their decision making (see Chapter 2);
- have ultimate responsibility for the long-term prosperity of the organisation;
- are accountable to their organisation's shareholders for overall performance.

In short, they deal with high-level and other strategy issues rather than getting involved in the detail. With respect to health and safety, this means that they probably do not want to know the organisation's approach to managing vehicle movements on its sites, just that they have one and whether it is any good or not.

## The tools

As explained previously, but it is worth repeating again, the tools that follow are based on experience which has shown that if they are chosen because they fit in the safety strategy and the organisational context, you can make great leaps.

### Board steering group

Most board meetings will contain an element of health and safety, even if it is just as simple as seeing the organisation's safety performance for the period. For many years, safety professionals have argued that this is not sufficient and it is common for many organisations to have a board steering group; such an approach has also been endorsed by various professional and trade bodies.

The idea of the steering group is that it is a subset of the main board and has the same standing and similar terms of reference as other 'off-shoots'. The steering group's aim is to maintain impetus at the senior levels of the organisation on health and safety matters.

As each organisation is set up differently, it is difficult to suggest who exactly should sit on the steering group. That said, the following (or their equivalents) are recommended along with a safety professional:

- operations director (chair);
- HR director;
- sales director;
- financial controller;
- property manager;
- regional manager ($\times$ 2).

*Operations director*

Typically, the operations director is responsible for the money-making side of the business and, by definition, the part of the business with the most people in as well as the side that has the most health and safety risk.

*HR director*

Generally speaking, it is the HR director that most organisations will appoint to be the director responsible for health and safety, and for good reason: they are independent from the operational line, hold a leadership position, and health and safety in the workplace should form an integral part of the organisation's people agenda. Business leaders with this sort of responsibility would not want to find themselves compromised by not being involved with the steering group.

*Sales director*

A strange addition one might think to a steering group aimed at health and safety is a sales director but consider, though, that the sales team are the ones who go out and win business. Is health and safety taken into account when setting up the pricing structure for the contract? Or is it operational colleagues who have to correct the problem afterwards while working within the margins agreed by the sales team. There is also another powerful reason for the sales director to be involved, not just because they are 'in charge' of people but because of the competitive advantage that can be levered by excelling in safety.

*Financial controller*

Finance tends to be another function within an organisation that is forgotten in relation to health and safety management, but at this level some buy in from finance is important to help them see the bigger picture of work involved and to help grease the wheels of the expenditure authorisation process.

*Property manager*

Many health and safety issues come down to building- and equipment-related issues. On occasions, it is difficult to get visibility of this subject to board members; the steering group provides that opportunity.

*Regional manager*

The steering group has a lot of potential to advance the safety culture of an organisation, but it can run the danger of having 'ivory tower' syndrome, so involvement of no more than two regional managers can help rein things back. The regional managers involved will undoubtedly start to be safety champions among their peers when they see just how seriously safety is taken at the top. Thinking back to Chapter 17, ideally these would be your early adopters. It is also a good opportunity for succession planning and development by giving them exposure to a different part of the organisation.

### Safety professional

The role of the safety professional in this steering group is really similar to the one they perform at a safety committee, only at a much higher level and thus you need to be polished. One of the many advantages the steering group poses for the safety professional is the opportunity to get business leaders and other influential people in the organisation to champion your cause.

### Other members

Depending on your organisation, there are other people you might want to include, although it is important that it does not turn into a 'committee'.

- **Senior representative from the insurers**  It is nice to get the nod from an external source to say 'Yes, I think you are going in the right direction' and, if approached, most employer's liability insurers will provide someone with the relevant background to do this. Not only does having this person on the steering group provide some assurances to the organisation that its approach is correct, but it enables the opportunity to learn about best practice from the insurer's other clients, as well as allowing the insurer to gain feedback as to how the insured is managing its risks.
- **A representative from the regulator**  Some would argue that having a regulator at such a meeting is a dangerous strategy. However, if the relationship is right, it can be hugely powerful to include them as, like having someone there from the insurers, they can see if you are going in the right direction as well as helping them influence the agenda. It is also helpful sometimes for someone other than the safety professional to challenge approaches. Depending on the industry is the organisation in this can be easy. However, for those in less regulated sectors, if you a primary authority relationship this can work in your favour.
- **Front-line colleagues**  It is easy to find yourself in steering group meetings talking about things that happen to people who are not represented or decisions about priorities and policies are agreed by people who are far removed from where the action takes place. Having a number of front-line colleagues involved in such a steering group provides the meeting with a conscience – in other words, they keep you honest.

Once the steering group has been established, they need some terms of reference so that they have something to talk about. Thinking back to Chapter 3, this sort of meeting is also really helpful to be part of the 'control' element of your safety strategy. As examples they could also:

- review the previous period's safety performance;
- review and agree the next period's improvement activities;
- identify existing road blocks to improving safety performance and remove;
- support the development of the safety strategy;
- update, and champion safety issues at, the main board meeting.

How often the steering group meets will generally depend on how much there is to talk about. However, given it is aimed at the high-level issues, once a quarter is normally adequate. That said, an emergency meeting may be required, for example, to discuss a prohibition notice being issued or to form a major accident review panel that looks at the outcome of a serious incident investigation. It is important to maintain momentum and this can be done by bringing in different external speakers to address the steering group. It is important, though, to align the speakers to the rest of the agenda. For example, a safety professional, following a review of the organisation's collision data, wanted to launch a driver safety training programme in the next six months and knew that it would require a certain amount of up front expense before the benefits could be realised. They arranged for an external speaker to attend one of the steering group meetings to help talk about the financial benefits of investing in this type of programme.

Pleasingly, it did the job as the meeting got fired up about it, not only because they could see the medium-term financial savings but they could also see they were exposed as an organisation and it was the right thing to do for their people. Needless to say, pushing through the funding for the programme after that became a lot easier, as did maintaining the momentum to get the programme up and running.

### High-level safety performance

There are many different ways to report safety performance and each organisation will have developed its own tools to do so. The problem with providing the board with safety performance is that either there are too many figures that mean little to them, or too much detail is given which does not get read. Obviously, finding the happy medium is important, but it can also be challenging. Of course, let's not forget that what we are talking about here is people, with families and friends, all of whom are affected when an organisation has poor safety performance or culture. It is therefore important to find a way to keep it personal as well.

It is often far better to provide a small number of measures that people really understand and can relate to. Therefore, experience suggests keeping three key performance indicators:

- accident frequency rate;
- severity rate;
- audit results.

### Accident frequency rate

This is a measure of how many lost time accidents are occurring per million hours worked and help with working out this rate is widely available. Using such data in this rate enables you to compare against other organisations' performance more easily, irrespective of their size.

### Severity rate

This is a measure of how bad accidents are, as it measures the time that injured parties have off as a result. The higher the rate, the more time off is taken, and thus the higher the cost to the business.

### Audit results

Chapter 7 outlined the concept of an operational high-risk audit. To balance the reactive data above, the board should get proactive data as well to allow them to see the whole picture. In this case, it may be the number of 'nos' given for the period and the year to date. Ultimately, the rates noted above and the results of the operational high-risk audit will provide enough detail for the board's purposes because all they need to know is how well the risks associated with the organisation are being managed; if they have any questions, or should they want extra clarification they will tell you.

### Making things more personal

To make things a little more personal, it is often useful to call the injured person by their name in reports and presentations rather than being more clinical and referring to them as the IP.

### League tables

Introducing competition between functions or operational divisions by comparing and ranking their performance is a concept that many organisations do, but there are some words of caution to remember:

- it can begin to foster a culture of under-reporting and thus it will become increasingly important to monitor reporting rates;
- regional managers can start to argue about every accident, which detracts your attention from other areas of your work.

## HSMS ownership

As mentioned previously, organisations normally nominate one business leader with responsibility for ensuring the implementation of the health and safety policy, which may or may not allow other members of the board to rest on their laurels. There are examples where organisations have gone one step further and assigned various parts of their HSMS to other leaders or senior managers. Figure 19.1 takes this and suggests some examples of which HSMSs could be owned by which leader.

For this type of approach to work, there has to be a willingness to make it work by the managing director. HSMS ownership needs to be aligned with each business leader's area of responsibility with a highly visible safety team to support it. This approach does not negate the need for a safety professional. In fact, it raises the profile of them immensely because not only do they write the policies (with other relevant stakeholders), but to make the approach successful they have to work with each leader in turn, something that may not happen with less contemporary thinking. Arguably, this is not for the faint-hearted and should be used when the senior management commitment is there and the organisation has really started to show characteristics of an advanced, positive safety culture.

| HR director | Operations director |
|---|---|
| • Misuse of drugs & alcohol<br>• Shift working<br>• Young persons<br>• New & expectant mothers<br>• Disability<br>• Stress | • Risk assessment<br>• Workplace transport safety<br>• COSHH<br>• Electricity<br>• Access & egress<br>• Fire |
| Finance director * | Marketing director |
| • Contractor management<br>• Legionnaires disease | • Communication<br>• Signs and signals |
| IT director | Sales director |
| • Display screen equipment | • Occupational road risk |

Figure 19.1  Examples of HSMS and business leader ownership
  * Where the property function report into finance.

## Cost control

Business is based on the principle of an organisation selling goods or services to another at a higher rate than it costs them to sell or provide them. The 'rate' is called the margin and organisations are always looking at ways to widen their margins, which can be done through sourcing cheaper products or running a more efficient operation with a smaller cost base. Health and safety is often seen as a function that wants to spend money. However, there are opportunities for safety professionals to help their organisation to widen their margins through providing opportunities to deliver cost savings.

In Chapter 8 we discussed how you can prepare a sound business case to support the implementation of safety improvement programmes. It is possible to take part of this approach and translate the cost of poor safety into a monetary value that can be measured periodically by the organisation's leadership. Using an example, we can explore this further. An organisation reported 80 lost time incidents in a year which led to 425 lost working days. Using the information already gained from previously preparing a business case, the safety professional knew that this meant there had been £1.5m worth (direct and indirect) costs as a result of accidents. During the year, the organisation was achieving a 10 per cent profit margin. This meant that they would have to have generated £15 million worth of sales to pay for the accidents alone. When the safety professional shared this figure with the organisation's leadership team, they realised that there was a big opportunity in improving safety and reducing costs at the same time. They decided that they would add this measure into their six-monthly safety performance review. The safety professional also used the measure to help them justify the improvement part of their annual budget.

## Managing costs

Identifying the scale of the financial benefits to improving safety performance is important as it helps to get people's attention, but there are other ways in which the safety professional can help widen the gap. The following are best practice examples linked to managing insurance-related costs.

### Claims modelling

Most, if not all, organisations will have an excess on their public and employers liability insurance, this is also known as the self-insured element. This can vary greatly and, depending on how much it is, organisations can end up paying vast amounts of money to claimants if the organisation is experiencing numerous claims which individually fall below the excess. This is, of course, on top of the existing premiums.

Claims modelling is a well-used technique of analysing the organisation's claims experience to determine what type of accidents are likely to result in a successful claim. Experience shows that when the claims experience is interrogated, the trends will centre upon the claimant's workplace, the amount of time off, whether any witnesses were present and the injury type. Once this information is known, all accident reports that come in are compared to the criteria and if they match they are rigorously investigated, irrespective of the accident's outcome. This creates two distinct advantages:

- a file can be compiled with a full incident investigation undertaken by a safety professional including photos, witness statements and the like taken as close to the time of the accident as possible. In other words, it overcomes lots of the common reasons why claims are sometimes difficult to defend.
- If the person involved in the incident is 'trying it on' they are less likely to consider putting a claim in if they think the organisation has the evidence necessary to defend itself. This does not affect those genuinely injured as a result of their employer's negligence who quite rightly deserve just recompense. If anything, it will help them.

### Recharging

Many organisations place an internal charge on a site's profit and loss account for any successful health and safety related claim. While this charge is unlikely to even cover the administrative cost of the claim, it does help focus the site manager's attention on the need to manage health and safety effectively as it 'hurts them in the pocket', particularly if their bonus is related to profit and loss performance while keeping the money in the organisation.

### Premium contribution

Another successful technique used by a number of organisations is to apportion insurance premium contribution based on the parts of the organisation that are having the most claims instead of it coming out of the central pot where either nobody feels the pinch or everyone pays the same. Again, this approach helps to keep people focused on the need to manage health and safety effectively. It may not be immediately obvious why the issue of recharging and premium contribution are matters for business leader engagement. However, given that they are likely to be radical changes from the existing arrangements and are likely to have far-reaching ramifications, authority from the most senior levels is required.

## External recognition

As discussed in Chapter 1, health and safety does not have to be all doom and gloom. Getting recognised for managing health and safety effectively is something worth shouting about and plenty of opportunities exist. Organisations can be recognised externally by independent bodies for their safety performance or the programmes they have implemented. This sort of recognition is great not just for the awards dinner and a director standing up and accepting the award, but they are also another example of how the safety professional can add value to the business by helping to deliver competitive advantage.

## Risk profiling

Risk and exposure to it is a key concern for business leaders and senior managers, indeed they are well versed in it particularly from an enterprise and general business risk point of view. Often safety will feature in there somewhere but this tends to be a one line item in the organisation's risk register. More enlightened organisations are expanding this to develop a more detailed safety risk profile enabling their business leaders and senior managers to see where they are exposed, so helping provide them with the chance to address it as well as drive performance improvements. Developing this sort of tool is not hard and while it works particularly well for multi-site organisations, it can be adjusted so that it makes distinctions between the risk posed by different departments on single site set-ups. As we discussed in Chapter 14, such tools are based on a mix of measures both objective and subjective, as well as leading and lagging. These in turn create a rating which can be aligned to a risk level – either high, medium or low.

Again, we can use an example to illustrate how this can be used. The leaders of a multi-site organisation wanted to improve their safety performance and wanted to be strategic in how they did it as well as spending their money effectively. The safety professional developed a risk profile tool that allowed the organisation to rank all of its sites from high to low risk. The leaders of the organisation then used this to work at the right level to drive real improvements in several ways:

- To help confirm that they were investing money in the right areas – e.g. the sites that had the most safety risk, providing additional assurance to the non-executive directors.
- To set their team's personal objectives to improve the overall risk profile of their particular part of the organisation as the tool showed them where they should be focusing their attention to get the best return on their time.

## Looking externally

It is foolhardy to think that either you have all the answers or you are the only ones having the problem you are experiencing. At business leader or senior manager level, the opportunity to look externally for answers can be a tricky thing to do. However, establishing networking or round table discussions on a subject is really effective.

To explain this more fully, let's use another example. A safety professional was working with the organisation's operations director and they both wanted to look outside of the organisation for inspiration to a safety problem they had. The safety professional made contact with their counterpart in another business and arranged for a meeting to be set up between them and their two operational leads.

Each organisation explained a little about their set-up and the challenges they faced; they also had a conversation about how they were addressing them. Over lunch, the two operations directors talked about the wider business issues they faced, and it became obvious that both organisations were facing similar challenges and were overcoming them in different ways, which each could benefit from. Within six months, both organisations' leadership teams had had round-table discussions about the three biggest challenges they shared and how each was dealing with it and, as before, the benefits of sharing ideas was clear to see.

## Leadership in action

So far, the tools we have looked at have been about things that business leaders can do in the 'office'. However, it is important to remember that their role as a figurehead is vital, too, giving them something to be the figurehead of in terms of safety is really powerful. As an example: an organisation had just launched a new near-miss reporting tool as they wanted to make more informed safety decisions based on this data. As part of embedding near-miss reporting into the organisation, the safety professional requested the managing director to ask that every time the business leaders and members of the senior management team went out to visit a site (irrespective of the reason for their visit), they always had a conversation with people about near-miss reporting – why it was important, how the information would be used and how to report them. When they were out with people on the front line, if they saw a near-miss they had to stop the team from working and have a chat about what a near-miss is, why it is important to flag them up and show them how to report the near-miss on their work phone. This approach went on for several months and over a twelve-month period the organisation saw an increase in near-miss reports from the low hundreds to over 7,000 (more than one per person in the organisation) by no other means than through leadership in action.

## Conclusion

As a population of the workforce, business leaders and senior managers, while they probably make up the smallest proportion, are arguably the most influential in any organisation. Therefore, the importance of engaging them in a meaningful way cannot be made clear enough. When considering how to get them involved and lead health and safety, you need to remember that they are extremely busy people. The tools discussed in this chapter are far from being complex or difficult, but what they do require is business leader and senior management involvement in order make them successful; and, of course, they provide great ways for them to demonstrate tangible things they are doing to lead health and safety within the organisation.

## Reference

1   www.iod.com/intershoproot/eCS/Store/en/pdfs/managers.pdf (accessed 14 November 2015).

# Chapter 20

# Middle and junior managers

In the previous chapter we discussed various methods that can be used to engage business leaders and senior managers to help sustainably improve an organisation's safety performance and culture. One thing that is certain is that once this group of people have become 'switched on' to it, they will push the wave through the organisation. This wave will invariably 'hit' their direct reports first, but so does everything else on their agenda whether it is sales, working capital, margin, achieving customer service level agreements or employee retention rates. However, the pressure does not stop there. They also get it from their team and others as well.

In this chapter we will explore some very simple, yet experience shows highly effective, 'game-changing' ways to get this population engaged in line with their roles and commitments.

## The role of middle and junior managers

As we did previously in order to identify the most effective tools for these people to use, it is important to understand more about the role that they fulfil in an organisation. The first thing to note here is that it is important to think of leadership and management as a scale, with each at opposite ends and having widely accepted and distinct aspects within them, as shown in Figure 20.1.

While a business leader or senior manager might spend more of their time towards the leadership end of the scale, they will inevitably end up doing some management-type activity, even if that is just making sure that the right process is followed for the approval of capital expenditure. The same is true for middle and junior managers, except that they will spend more time towards the management end of the scale, but will end up at some point during their day playing more of a leadership role.

With this in mind, we can start to think about the game-changing ways in which we can engage this population with safety, but it is really important to remember that you are competing with many other areas, all bidding for their attention, each thinking their area is more important than the next. In

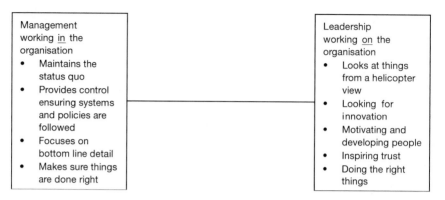

*Figure 20.1* Leadership and management roles as a scale with widely accepted and distinct elements of each

reality, the middle manager's job is not an easy one; they have to deliver the performance the board want in challenging circumstances where no two locations or teams they are responsible for are likely to be the same. For these reasons, middle management is arguably one of the hardest groups of the organisational population to engage.

## The tools

Successfully engaging middle and junior levels of management in order to drive safety performance and cultural change comes down to having their boss on side, breaking things down into 'penny packets', making things snappy and to build on existing systems and processes – coupled, of course, with the elements we have already discussed in other parts of this book, as whatever you choose has to fit your organisation and the people in it.

### Safety development

While safety learning and development is important for any role, often middle and junior management tend to get forgotten as the safety professionals in the organisation either focus on front-line colleagues and those in more traditional leadership and senior management positions. Before you start running off and developing a training course for your middle manager population, refer back to the principles we discussed in Chapter 5 to work out exactly what it is you want to achieve as this will help determine the route you take.

To illustrate this we can use the following example. An organisation wanted to drive more line management accountability for safety, specifically aiming at their area manager population. As part of the strategy the safety team was reduced and the services they provided changed. At the same time all the area managers were put through a formal entry-level safety qualification. When they had successfully completed this they became the focus for their teams safety related queries whereas before the team would bypass their line manager and go straight to the safety adviser for help. Of course, the safety team were still on hand to help if things were complex or became too big.

Whether you agree with this sort of approach or not is irrelevant as it was right for this organisation at the time. The bigger point to consider is how can training and development for middle and junior managers, who have many plates to spin, help drive sustainable improvements in the organisation's safety performance and culture. There are numerous development options available – the crucial thing is to work out what you want this management population to do differently, what is the game-changing behaviour you want them to show and then when you have done this, you can work out what the training should be.

### Performance-related objectives

It has become common for organisations to give those in management roles performance-related objectives which tend to be bonus-worthy. Their aim is simple: to motivate people to perform at the highest level they can and, in return, they get some form of reward – normally cash – while the organisation reaps the benefits of their work – everybody wins. Their use is based on well-known and recognised psychological theories related to positive and negative reinforcement and, of course, for most people the primary reason they go to work is to earn money to enable them to have a life outside of it; therefore, incentivising them to deliver something outside of the norm or 'raise the bar' by stretching their current performance can be a good thing to do.

There are two main types of objective:

- **Outcome** These focus the person on delivering a numbered outcome to which their performance is measured – for example, a reduction in lost working days associated with accidents at work or an increase in near-miss reports.
- **Output** These focus the person on delivering an activity or programme that is linked to the other overall safety strategy. Their success at this can be measured in a number of ways: the content of the programme, feedback from those it affects on its relevance, and so on.

#### Outcome safety objectives

Careful consideration has to be given to using this type of objective because, on the face of it, it has great potential. Yet if it is not done properly it can have the complete opposite effect of what you are trying to do, as the following example shows.

An organisation set a bonus-worthy objective for all its regional managers to reduce their accident frequency rate by 10 per cent compared to the previous year. Some of the managers approached this in the way the safety professional had envisaged when he or she wrote it, by trying to make sure that safety was managed more effectively in their region. Sadly, others did not and their behaviour when a lost time accident was reported drove under-reporting in their business. In other words, the organisation's safety performance looked as though it had improved when it was not the case.

If you do decide that this approach is the right way for your organisation, make sure that you have in place sound governance – checks and balances – to make sure that the results are real and that you are not driving the wrong behaviours. A simple way to do this is to look at the relationship between more serious accidents and other types; these are often explained by the way of an 'accident triangle'.

That said, it is possible to set an outcome objective that is more likely to drive the right behaviours based on a proactive measure as the following example shows. Over recent years, an organisation's safety audit scores had been slipping. As part of the improvement plan to tackle this they decided to set a bonus-worthy outcome objective for all their area managers of increasing their branches' audit scores by at least one point (the scoring system was from one to ten). Initially, the safety professional picked up that a number of the area managers were simply challenging the auditors whenever they issued a report by arguing the non-compliances they had identified to try to get them to review their score. Within two months, the managers realised that they were unlikely to change the auditor's mind, so the safety professional stepped in to give them a framework to work with. At their management meeting they talked through the common audit failures over the last twelve months and how they can be addressed (as they were doing so, the group noted that many of the failures were really easy to put right) and the group agreed they should take some action over locations that did not achieve an increase on their previous score. Over the remainder of the year, the audit scores steadily increased.

In the main, outcome objectives are better suited to measuring the overall effectiveness of safety strategies. However, they can be used effectively to measure individuals' performance. provided that they focus on the right measure of success, which in the main need to be proactive and not influenced by factors beyond the person's control or that of their team.

### Output safety objectives

These tend to be better suited to the management population as they focus people's minds on working on the things that will enable the safety strategy to be delivered as the following example shows. An organisation wanted to drive more line-management accountability for safety, specifically aiming at their middle-management population. They set each member of this group a bonus-worthy output objective related to the delivery of one of the actions in their safety plan. As this was quite a change from what the organisation had done previously, the safety professional noted that the level of interest from this group of managers increased as they each were in contact to understand more about the objective and what they needed to do.

There are several advantages to this type of approach:

- With each member of the population leading an area of the safety plan, part of the success criteria should be that they test whatever they come up with to fulfil the objective in their part of the organisation first. Thinking back to Chapter 17, this means that they become the innovators and their team the early adopters – in other words, the cultural change happens faster and 'in team'.
- When they have successfully developed and tested whatever they have come up with to achieve the objective, they champion its roll-out among their peers. This means that acceptance of the approach and roll-out of it across the organisation is easier and faster as experience shows peer-to-peer influencing is more effective in getting things done than someone perceived to be from outside the team telling them they should do it.
- The role of the safety professional in this approach is to work with the managers to help them shape the output, not to do it for them but to use coaching in order for them to deliver. This places the safety professional on a different footing to the more traditional view of them that we have already explored, and it is another way to help overcome any challenges there might be with resource.

The only word of caution with this approach is that it will only work if you have successfully got their managers on side to start with, as not only are they the ones who have to formally write the objectives, but if the management team do not like the change they will simply go to their manager to seek to change it.

### Working at the right level with incident investigations

All too often experience suggests that the details of accidents go below the radar of middle and junior managers. Yet the power of this management population talking to the injured person about their accident and having a challenging conversation with their direct reports about what they are doing to prevent a similar reoccurrence should not be underestimated. This is not about them doing an accident investigation as many organisations ask them to do. The injured person's line manager should do this and the middle manager should simply evaluate what they have done and seek to get it independently investigated if there is an apparent conflict of interest. The approach outlined here begins to become a game changer in terms of positively affecting the safety culture for the following reasons:

- It drives the middle manager to work at the right level, reducing the amount of perceived effort they have put in while reinforcing the view that safety is a line management responsibility.
- It improves the quality of incident investigation and level of learning that can be achieved as it drives front-line managers to adopt a mindset of 'I'd better do this properly because my boss will want to know what I've done'. Through repeated exposure over time, their mindset will move towards one of 'I want to stop this from happening again', which is a significant step in positive safety culture change.
- It connects the injured person with their second line manager. Experience shows to front-line colleagues that this person is the 'big boss' as they are senior to their immediate line manager and more visible to them than a business leader or senior manager. Middle managers showing genuine concern for the person and being inquisitive about what happened helps to inspire trust and drives the middle manager to really understand what is happening in their part of the organisation.

### Peer reviews

As we have said before, a peer talking to a peer is a really powerful conversation although most do not realise it. Experience shows that people tend to be more accepting of someone who works with them in their team as an equal. Challenging them in a conversational tone is different and it feels 'safer' compared to when they have this type of conversation with their line manager or with an outsider to the team. From a safety culture perspective, and those wider cultural issues, this is the area where we need to be. Peer-to-peer reviews can be used effectively either for reactive or proactive measures as the following examples outline.

#### Reactive peer reviews

A large single site organisation wanted to spread best practice and start to highlight to their management team that there were common underlying issues across the site that were affecting safety and, irrespective of the specific part of the plant they were responsible for, they were affected by the same things. The safety professional arranged for the site's general manager to set up a weekly safety review meeting with all the departmental managers in attendance, although to enable a more free-flowing conversation the general manager would not attend and it would be facilitated by the safety professional. At each meeting the departmental manager discussed the accidents and near-misses that had been reported the previous week and the learning points. Other members of the meeting were then given the opportunity to delve deeper into the incidents and challenge the conclusions reached. To start with, the level of engagement was slow, but using the ideas associated with the law of diffusion of innovation (discussed in Chapter 17) they worked with two people they thought would be early adopters and encouraged them privately to be more challenging in the meeting. In time, this happened and more people got involved. The meeting became more productive with the outputs being used at three levels:

- Each manager took the best practice back to their teams to discuss and adopt.
- The key themes were collated and shared with the site's senior management team at the monthly site management meeting.
- The themes were used to test to see if the site's safety plan was covering the right areas and were also added to the list of things to do.

### Proactive peer-to-peer reviews

An organisation wanted to drive performance improvements through providing a fresh pair of eyes on how safety was being managed in each region. To do this the safety professional arranged for each regional manager to visit another region to review what they had in place, when they were there they considered the following two questions:

- What have you seen in the region that you intend to take back to yours to help improve your safety performance and culture?
- If you were the regional manager of this region, what would you change to help improve your safety performance and culture?

Having completed this exercise, the regional managers fedback to their counterpart, feedback from those taking part was really positive with plenty of evidence of them putting into place both what they had seen elsewhere and what their peers had suggested. This type of approach can, of course, be used for front-line managers to help them drive forward their safety performance and culture. At this level, however, it is useful to provide more detail in the questions that you ask them to look at.

### Performance management

Every manager should be aware of the concept of performance management; it is about the line manager ensuring that the results are delivered by their team and it works at all levels of an organisation. The mechanisms for performance management are generally well established, yet many organisations and safety professionals struggle to make safety a performance management issue. However, if you can overcome this, for some organisations, real benefits can be realised, as the following example outlines.

A large multi-site organisation refreshed and launched their risk assessments and wanted to make sure that the site managers where fulfilling their requirements of the roll-out. The approach to risk assessment they had decided on was to use a system of 'model' risk assessments which are based on worst-case scenarios, and site managers are expected to review each assessment in turn on a calendarised basis to make site specific as necessary and ensure the assessment's findings are implemented.

The process for ensuring that this was undertaken correctly and in a timely manner was built into the existing performance management system so that the area manager visited each of their sites on their monthly visit; part of the discussion with the manager was about health and safety. Specifically, this conversation would cover where the site manager was with implementing the assessments. If they were on track, fine; if not, the area manager offered support to get the site manager to improve their performance or if they were repeatedly 'off course', further steps were taken within the organisation's performance management and disciplinary framework.

The safety professional also arranged for this approach to replicate at the next level up by the area manager's director so that it became a discussion point at their one-to-one meetings. They also put some checks and balances in place to provide some assurance that things were being done and they linked this to the operational high risk review discussed in Chapter 7. This approach is shown in Figure 20.2, the outputs of which were that undertaking or reviewing risk assessments did not become an annual event. It became much more regular and the follow-up conversations between the area manager and their site managers, and the area managers and their director ensured that it stayed on the radar, so that in time this became how the organisation does risk assessment.

This approach, of course, works for just about any safety management topic or improvement programme. While we can use the law of diffusion of innovation to help create a pull for things from people, sometimes you need to manage performance too. Managing performance and maintaining control to a standard is, of course, key to a manager's role.

### Safety conversations

While middle and junior managers tend to spend more time towards the management end of the scale shown in Figure 20.1, they fulfil a leadership role as well. Thinking back to the things that leaders do, some of the most important where people are concerned are in relation to motivation

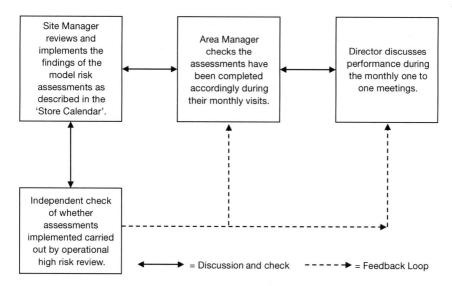

*Figure 20.2* A simplified health and safety performance management system in schematic form

and inspiring trust, things that are really important when we talk about significantly and sustainably improving safety performance and culture.

More enlightened organisations have built on the approach outlined in Chapter 7, so that when members of the middle management population go out and spend time with their front-line colleagues, they have more engaging conversations that inspire trust and motivate. In other words, they can start to address those issues that are under the water-line, rather than simply complete a tick-box safety inspection, which is common in many organisations.

Safety conversations are generally grouped into three: those where the purpose of them is to:

- motivate those the manager is talking with;
- assure the manager that process and policies are being followed;
- educate either the manager or those they are talking with about something.

Depending on the type of conversation the manager is planning to have dictates the sorts of questions they need to use in order to get people talking and, just as importantly, start to capture some of the undercurrent issues. Table 20.1 outlines some key questions.

The key to success with safety conversations is to be clear about what it is you are going to talk about as well as really listening to the answers being given, as the following example shows. An organisation had recently rolled

*Table 20.1* Typical safety conversation questions

| Type of safety conversation | Typical questions |
| --- | --- |
| Educational | What work are you involved in here?<br>What could be done to make work safer?<br>What are the key dangers in this work?<br>If you have concerns or ideas about safety, how do you raise them?<br>How well are they listened to?<br>How are work priorities communicated to you?<br>What pressures are you under?<br>What helps or hinders you in the way we go about it?<br>What should we (leadership) be doing differently? |
| Motivational | What is the worst (kind of accident) that could happen?<br>What is in place to prevent it?<br>How confident are you in these protective measures?<br>How effective are they?<br>How could they be improved?<br>How easy are they to follow?<br>What would be the safest way to do it?<br>How do you know it's safe?<br>What systems and procedures apply? |
| Assurance | How could you get hurt (or killed)?<br>What would help to make you safer?<br>If there was one thing we could do to reduce the risk?<br>What are you and the others most concerned about?<br>What are the obstacles to complying with safety rules?<br>How would we (leadership) help? |

out a new piece of equipment to help improve the safety of front-line colleagues. Part of the roll-out plan involved creating some good news stories about how the equipment was being used and what colleagues thought of it to help others who were, in law of the diffusion of innovation speak, the late majority and lagger groups accept the change to their ways of working. As well as this, the safety professional and operations director decided that the area manager team, when they were visiting branches, should spend some time having motivational style safety conversations to help promote the equipment further. The safety professional provided some development for the area managers at their next meeting about safety conversations, the type of questions to ask, how to be an active listener and how to give praise when people get things right.

As part of the three-month post-implementation review of the new equipment, the area managers provided feedback on the outcome of their safety conversations and explained that:

- Where they had met resistance towards the equipment, despite their motivational approach, they arranged for people from a neighbouring branch who liked the equipment and used it to spend time with their peers at the branch that did not – they were surprised just how well this approach worked.
- A theme was emerging that there were some issues with the equipment that had not been picked up before, and if they had not gone out and talked to people about the equipment they were not convinced they would have come to light until it was too late.

This example shows two very important things about safety conversations. In their first piece of feedback the approach the area managers took was about inspiring trust that the equipment worked by saying, in effect, 'OK, let me prove to you that it works . . .' and inspiring trust is an important aspect of leadership.

The second point is that although you might go in with the intention of having one type of conversation, it can very easily morph into another. When this happens, in the main it is an indicator that you are having a really rich conversation. Indeed, the second learning point the area managers highlighted in this example was more from an assurance type of conversation, yet by their own admission if they had not gone out and had the conversation, the issue might not have been identified for some time.

Experience shows that some managers feel uncomfortable with this approach as it is quite removed from the more traditional approach to command and control leadership. However, the positive impact it can have on safety outweighs this, provided the tools are used at the right time for your organisation. Of course, safety conversations should not be limited to just middle and junior managers. The approach is just as effective if it is followed by business leaders and senior managers as well as those managers on the front line.

### Safety as a selection tool

Appointing someone into a role is one of the most important things you can do as a manager, as you are letting someone come into the organisation and your team as well as expecting them to deliver for you. If you make a bad recruitment decision, it can cost you dearly to performance manage the person as well as the hit on team morale and overall organisational performance. With this in mind, it is little surprise that organisations have well-established recruitment and selection practices to help select the best candidate for the job. Experience suggests that few of those selection techniques include elements on health and safety, which is odd as experience suggests that how good you are at managing health and safety is a good indicator of how effective the person is at general management.

It is widely acknowledged that safety culture is driven by several factors including by the local manager. Given how organisations are set up, middle and junior managers tend to be the ones who recruit such individuals more regularly as the volume of these roles means there is more churn. Including safety-related questions in an interview process, or an exercise in an assessment centre, is really powerful.

Examples of such questions that can be used are shown in Table 20.2, along with a scoring criteria for organisations who use a more rigid selection process. It is important to say that you cannot expect the answers given to these questions to be the only ones considered when making recruitment decisions, but:

• they set the tone that safety is important to the organisation, so much so that it is part of the process to get a job;
• it is another indicator of a candidate's attitude, the way they tackle problems and engage their people;
• they provide an early indicator to the recruiting manager of their possible safety development areas, so they are easier to plan for.

To illustrate this, another worked example can be given. An area manager was recruiting for a new branch manager's position and included some safety questions in the process. Of the final two candidates, one was an internal existing branch manager and one was external to the organisation, and the area manager was torn between who should get the role. In the end the manager reviewed the answers given in the candidates' interviews and decided to give the bigger branch to the internal candidate as there were a number of safety challenges that needed to be addressed there, and they demonstrated through their responses that they had a good track record of managing safety in the organisation's environment and could engage people well. This created another vacancy that the external candidate was offered as it was easier to manage and would be a good opportunity for them to learn how the organisation managed safety before they embarked on managing a less unforgiving branch.

## Conclusion

Middle and junior managers in an organisation have a really tough role as they are expected to deliver a range of things and often everything has to go through them in order to get to the front line. Delivering significantly improving safety performance and cultural change using this organisational population is about providing them with quick and easy tools that align with their role as managers so that they work at the right level.

Table 20.2 An example of safety-related recruitment questions and scoring criteria

| Question | Good response | Average response | Poor response |
|---|---|---|---|
| Tell us about a health and safety problem that you have faced at work and how you dealt with it. | The problem was well explained. They did not shy away from the problem and flagged it up straight away. They helped work with their managers/colleagues to develop a solution. | The problem was well explained. They reported the problem. They showed some attempt to work with their managers/colleagues to develop a solution. | The problem was not well explained. They reported the issue. They left it with their manager/someone else to resolve. |
| Give us an example of how you have managed to get colleagues to engage more with health and safety. | Example used was relevant and clearly explained. They took time to understand other people's points of view. Evidence that they provided to their people a personal compelling reason why safety is important to them. Showed an engagement strategy – e.g. they worked with the early adopters to build momentum rather than trying to change everyone's opinion at the same time. They explained whether their approach worked and what they learnt from it. | Example used was relevant and explained. Some time was taken to understand other people's points of view, but based their approach on what they thought. They tried to explain to their people a personal compelling reason why safety is important to them. Some thought had been given about how to approach engaging people although it was somewhat 'hit and miss' or lacked structure. They explained whether their approach worked. | Example used was relevant but not very well explained. Little time was taken to understand other people's points of view and their leadership style was more command-and-control. They did not provide a personal compelling reason why safety is important to them or, if they did, it was weak. No thought seemed to have been given about how to approach engaging people. It was unclear whether their approach worked. |

# Chapter 21

# Front-line managers and employees

Business leaders and middle managers might have a great deal of influence over an organisation's safety performance and culture, but they do not tend to be the ones that have the accidents nor are they the ones with the immediate management control over the risks that create them. So far, we have explored ways in which other organisational populations can help drive beneficial changes in safety performance and culture. In this chapter, though, we will seek to outline those tools that experience shows can help change front-line workers' and managers' perception of safety and, while incredibly simple, drive transformational change.

Of course, very traditional views about engaging front-line workers and managers centre around compliance and the argument that, as long as you provide safe plant and equipment for use and provided people follow the rules, accidents will not happen. While this might be a true statement, life is rarely such. Although there are legal requirements for workers and managers to follow the rules of safe working, you have to be quite unlucky to be prosecuted. With this in mind, a more visible 'deterrent' to employees working unsafely is the requirements of their contracts of employment, backed up by their organisation's disciplinary policy, which normally states that a breach of health and safety rules is a potential circumstance for summary dismissal. There is a place for these sorts of 'sticks' but there also needs to be some 'carrots' too. And if the right 'carrots' are used, they can really deliver sustainable improvements in safety performance and drive not only safety but wider cultural change.

## Front-line workers and managers

Experience suggests that there is a misconception among some organisations and safety professionals that people who work on the front line (workers and their immediate line managers) are not interested in safety – they simply come in, do their job and go home – and therefore we should make things as simple as we can for them and try to remove the need for them to think for themselves. Yet while they might want to just come in, get the job done and go home in the vast majority of cases, they are interested in safety and have very valid views on how things are done.

The first step in gaining employee ownership is to start dialogue between the organisation and the employees. In the UK this mechanism has long been established with its own legal provision, traditionally in the form of health and safety committees. The concept of safety committees is a good one and can really help to drive things forward. However, experience shows that there can be some drawbacks to them:

- They can end up in a talking shop where very few tangible actions are delivered.
- They can break down into a them (management) and us (workers) situation.
- They can fail to engage with everyone in the workplace.

## Tools

While a safety committee is a reasonable first step, to deliver transformational safety performance and cultural change you need to do much more to drive front-line workers' and managers' safety engagement. With this in mind, what follows are a number of tools that have proved, time after time, to deliver that sort of benefit, provided, as we have discussed previously, you pick the right ones (or adapt the right ones) for your organisation at the right time.

### Safety champion programme

When the concept of a safety committee was first formalised in legislation in the UK during the mid to late 1970s, they were built on the idea that the work-force would vote for their peers to represent them through their trade union. As trade unions have declined, other arrangements have been introduced to reflect this and still get workers' representatives to talk about safety with their employer. One of the main successes of these approaches is that they promote 'bottom-up' involvement and decision making – in other words, the people who do the job have a say in how it is done, not just those who sit round a board table. The safety champion programme builds on this principle.

Experience shows that nurturing people from the front line to become safety champions can have positive effects far beyond just safety. The aim is for the champions:

- to provide the front-line manager with another pair of hands for them to use to help them manage safety at their location;
- to provide feedback about new safety initiatives or problems;
- to act as early adopters to new ways of working or thinking.

Getting the right person to become a champion is critical and not just given to the person who looked up at the wrong time. As with all other roles, there should be a job description and person specification, and this one is no different. Experience suggests that the person specification should include things like:

- being reliable and trustworthy;
- a good problem solver and team player;
- being approachable;
- being interested in health and safety;
- being respected by their peers for working safely.

There are two major differences between the safety champion role and that of the normal 'job'. These need to be addressed from the start, as getting buy in from senior managers for this programme will not be obtained unless they are. These differences are that the safety champion programme is not going to increase the organisation's head count. In other words, this is an extension of the person's existing job – they are not the location's 'full-time health and safety person', that remains with the manager. The second is that they may not get paid any extra for doing the role.

Often, criticism of organisations is that they continually 'dump' things on front-line management and expect them to do it despite their existing workloads; clearly, this programme helps to overcome this, as the organisation can be seen as providing another pair of hands for the already busy location manager to manage safety in their workplace at little additional cost. Typically, there will also be resistance to the programme as people may think that it will take the safety champion away from their 'day job' for long periods of time and end up costing the organisation that way. However, in reality the safety champion's role is only likely to require the person to be doing 'non-production' type work for half an hour or so a week, which in the scheme of things is not much, particularly if it frees up some time for the manager to do other things.

While the safety champion may not get paid any more for performing these additional duties, there does need to be some form of incentive for them. Some organisations do this by giving them a little something extra in their salary, in much the same way as organisations give a premium to first aiders. Other selling points, however, are that:

- they will receive training (which, depending on the training course used, could be accredited by a professional body);
- the programme allows people to get on to the first rung of the safety professions career ladder;
- it provides an opportunity to highlight to the organisation their potential and talent outside their normal field of work;

- they can make a difference to improve workplace safety for themselves and their colleagues.

It would be extremely foolhardy and naive to put someone in post without giving them any training and, as discussed previously, there are many choices in this area. While formal courses will fit the bill in the majority of cases, if the budget does not allow, the organisation could deliver in-house training based on the specific tasks they require the safety champions to do and then build up to a more in-depth course when they have proved their worth, which they will.

Once the training has been completed, it is important to ensure that the safety champions have some direction so that they know what they are doing when they go back to the workplace. However, it would be unfair just to leave them to it. Thus, it is important that the safety professional keeps in touch with them and provides support when they need it.

Once there is more than one safety champion, it becomes possible to establish a safety champion network which can help support each other as well as coming together, on a biannual basis, to discuss wider safety programmes and challenges.

Without doubt, this programme works well in all workplaces, but there are ten important steps to go through to increase your chances of success:

- gain organisational support;
- determine budget constraints;
- agree a considered job description and person specification;
- recruitment and selection;
- identification of training provider;
- training;
- determine, in conjunction with the location manager, specific tasks for them to undertake in their workplace;
- introduce personal development plans;
- maintain regular contact;
- refresher training as necessary.

The following example shows how this approach can be used. A multi-site organisation identified that their location managers were struggling to maintain local safety management arrangements because their jobs had grown dramatically in recent years. At the same time their safety professional noticed that there were several people around the location network who were particularly interested in safety. To build on this and help address part of the wider management capacity issue, the organisation decided to introduce safety champions at their larger locations to support the manager to:

- undertake risk assessments and brief colleagues on the safe systems of work;

- help with incident investigations and encourage their peers to report near misses;
- give feedback to the manager regarding what safety arrangements were working well and where things could be improved.

The organisation decided to work with the people who had already showed an interest in this area as they identified they were their early adopters. As costs were being scrutinised and, as it was a pilot programme, the safety professional provided the early adopters with a course in risk assessment and incident investigation. When they left the session they all sat down with their line managers and agreed what they were going to take on when they got back to their location.

The safety professional checked in regularly with the champions and their managers to give them some encouragement to keep going and see how things were progressing. This carried on for six months. When the organisation came to review the success of the pilot programme they found that:

- Those locations that had been audited in the last six months had improved scores compared to those that did not have champions.
- Near-miss reports in those locations where their champions were was significantly higher than those without and crucially the people reporting the near-misses were not just the champion or the manager.
- Managers reported as having more time to monitor safety arrangements at the location and what the champion was doing, as well as having more confidence that things were being done right, enabling them to spend more time on other priority areas.

After this review the organisation decided to widen the programme out to the next group of people that had shown an interest (their early and late majority) and decided to invest a little more in their safety training with a formal qualification.

### What is our shared safety vision here?

Often with cultural change programmes they are devised centrally and pushed down through an organisation. In many cases this is useful. However, there is an argument to say that the best people to drive the change are those affected by it and that the role of the organisation is to provide some tools to help them tackle the below the waterline issues.

As we discussed when we looked at determining a safety strategy, you need to know where you are now and where you want to get to before you can work out how you can get there. A powerful tool for teams to do this themselves is to ask them to describe what safety used to be like (as this is a safer way for them to say what safety is like now) and what they would like it to be. Having done this, the team can start to work out what they need to do to get to the 'better place'.

To help illustrate this, we will use the following example. A safety professional and the operations director wanted to drive individual teams to improve their safety culture. Instead of telling them what to do they decided on a different approach. At the next operations meeting they took the branch managers through the type of session they wanted them to run with their teams. They asked them to individually answer three questions, thinking about this team and how we work:

- What did safety use to be like?
- What do we want safety to be like in the near future?
- What do we as a team need to do differently?

They then had a discussion about what they had come up with. After some discussion they agreed what they should be aiming for and the actions they are going to take to get there. Next they ran similar sessions with the teams back at their branches. Within a month all the branches had their own safety cultural improvement plan and they were surprised about the fact that when the managers compared where they wanted to get to, just how much similarity there was.

In order to keep this programme alive, every quarter the branch managers, as part of the team brief, process reviewed the actions that they had committed to see how far things have progressed. Twelve months after the original roll-out of the programme it became clear that each team had started to make a move towards the safety outcomes they wanted.

This programme has many benefits:

- While we often say that each organisation has a different culture, so too does every team within an organisation. Cultural change needs to happen within teams and therefore this programme helps provide a framework for teams to work independently while heading in the same direction.
- It tackles team-specific issues and helps them come up with approaches to tackle it themselves; of course, the safety professional needs to be on hand to support with the more tricky aspects.
- Team members can start to hold each other to account for the actions they do or don't do, as well as what they say or don't say.
- This is the first step towards a behavioural safety programme so it is easier for you to gauge whether such a full-blown programme is right for your organisation.

### Golden safety rules

All too often people say that they do not follow safety rules because they are 'too complicated', 'too hard to remember' or 'not workable'. For a number of years, more enlightened organisations have been working on the same principles that Loftstedt presented (see Chapter 1) and simplifying their key safety rules, and many more organisations are following suit. Call them what you will: Cardinal Rules, Golden Safety Rules or Lifesaving Rules, they are the absolute fundamental safety rules people need to follow in order to stay safe when working for a particular organisation.

There is, of course, a place for detailed risk assessments and safe systems of work, but they should be proportionate to the risks they are designed to address. There are many people who are frightened about giving up the perceived protection they have in their existing safety rules by simplifying them. And sadly it is a perception that they are protected because it is, at best, naive to think that everyone in every organisation right *now* is following every safety rule their firm has to protect its workers.

It is with this in mind that more enlightened organisations have decided to go with an 80–20 rule and developed a handful of rules to enable people to work safely 80 per cent of the time, leaving the more detailed rules for the remainder of the work. Taking this type of approach is really powerful and helps make life easier for people working in front-line roles. After all, who will argue against what you are saying, which is effectively: why wouldn't you follow a handful of rules?

In order to simplify your safety rules, there are a number of steps to follow:

- **Understand your risks**   To make sure you do not miss anything and maintain your credibility, it is important to design rules that relate to the organisation's real risks and not just to address a number of hazards (the difference is subtle but the effect can be massive). This can be from a blend of reviewing your risk data – e.g. accident and claims experience, and also the perception of your workers and your view as a safety professional.
- **Slimming down your rules**   This is best done starting afresh, so for now ignore your existing risk assessments and safe systems of work. Write down for each risk what the basic and critical safety requirements are. Having done this, you then compare it to your risk assessments and safe systems of work to see if you have missed anything 'big' out. The chances are, though, that you won't have, because the idea of these rules is that they are so blindingly obvious. This is not to say that all your risk assessments and safe systems of work are redundant, though. They sit in the background to help managers know what they should be doing and to give people the outline of how to work safely, but if they forget all that stuff so long as they remember the 'cardinal rules', they will still be safe.
- **Punishment**   There is no point having a rule if there is no penalty for breaking it. Breaking safety rules, in all but a few cases, tend to go

unrecognised or unpunished for a host of reasons. However, in order to get the most of the 'cardinal rules', you need to make it worthwhile for people to follow them. In many organisations this is a step change in how they deal with safety breaches, but it does not mean necessarily that the person will get disciplined or sacked. It is simply a way of 'ramping up' the amount of management focus placed on breaking safety rules. Done properly, such investigations can identify if someone needs more training in order to help them follow the rules, identify if their manager put them in a situation that meant they couldn't follow the rules (in which case the investigation would shift to the manager), or just help identify repeat offenders which the investigating manager should take into consideration.

- **Selling it to your workers**  Your 'cardinal rules' will be little more than a desktop exercise unless the management team and the people on the front line come with you on the journey. From experience, to help this happen you simply need to repeat the exercise you did to develop them in the first place.

To help bring this to life, we can use the following example of a multi-site organisation which, through their business leader and middle manager's safety tours, have picked up that people are getting confused about the safety rules so that there is inconsistent application of rules across the organisation. The safety professional was tasked to try to address this and, having spent time with front-line workers and manager, found that there was a real issue to address. They decided to use this opportunity to really advance the organisation's safety culture and performance by introducing a set of cardinal rules.

First, they went through the organisation's risk data and by having a number of focus groups with front-line workers, determined what the real risks are (the things that are most likely to kill or seriously injury someone working there). These were mainly in the areas of working at height and vehicular movements. As a final sense check they compared their results to the rest of their industry sector. They decided on the following as cardinal rules in relation to vehicle movements:

- Only drive a vehicle if you have the relevant valid driving licence.
- Always wear a seatbelt when operating a vehicle.
- Never leave a vehicle unattended with its operating keys in the ignition.
- Give way to pedestrians.
- When walking, always wear a yellow, high-visibility vest.

They then did the same for working at height. Next, they agreed with the leadership team that if anybody breaks one of the cardinal rules, they will be subject to a disciplinary investigation. Having determined the rules (using front-line worker and manager involvement) and sought the buy in from the leadership team the cardinal rules were rolled out. The safety professional approached this by going along to the operational team meetings and asking just three questions:

- What do you think are the activities that are most likely to kill or seriously injury someone here?
- What are the basic things we can do to prevent someone from being killed or injured as a result?
- What should happen if we break these rules?

They finished off by explaining where the organisation's risk assessments and safe systems of work fitted in. Those at the team meetings then went away to brief their teams in a similar way. The safety professional also did some of the more transactional promotion such as arranging for posters to be put up and for key-rings to be given out in order to get people in the organisation talking about the rules.

Given that this was a big departure away from what the organisation was used to, it was important to continue to embed the rules and maintain people's confidence in them. To do this, the safety professional asked for the leadership team's help so that they and the senior management team would focus on them when they were doing safety tours covering the following points:

- not just pointing out when people are not following them, but praising people when they are following them;
- getting front-line workers to talk to people about the rules and whether they really do cover the real risks where they work;
- exploring what difficulties they might face when they follow them.

The outcome of these conversations feed back into a wider review of the impact and benefit of the rules. Six months later it was fair to say that this approach had driven forward improvements to the organisation's safety culture and performance. It also delivered other advantages in two very distinct areas:

- **Removal of confusion**   There is one simple rule for doing the same job everywhere. Many multi-site organisations have different ways of working in each of their sites, which can present problems.
- **Enabled challenge**   If someone asks you to break a rule you have the right to say 'no' and nothing will happen to you as a result. This is really important where there might be a perceived 'blame culture'.

*Table 21.1* Basic dos and don'ts of a golden safety rule programme

| Do | Don't |
|---|---|
| Focus on the real risks to your workers' health and safety. | Have more than ten rules (if you can help it). |
| Take your people with you. | Tell people what the rules are; they need to feel they thought of them. |
| Decide what the consequences will be for not following the rules. | Think what you have now is perfect. |

The ultimate check was that in the eighteen months that followed the implementation of the rules, there were unfortunately a small number of serious accidents that, if the injured person had followed the rules, the chances are they would have avoided injury completely, or if they did suffer an injury they would not have been as seriously hurt as they were.

There is one final learning point that experience shows with this sort of programme: there are some definite dos and don'ts (shown in Table 21.1) that should be borne in mind in order to encourage their take-up.

### Safety forums with a difference

As we have said before, in organisations with a more traditional view towards health and safety, decisions are typically made in boardrooms and implemented often with little meaningful engagement with front-line workers and managers. One way to overcome this, which is simple to implement and has many benefits, is to introduce safety forums made up of representatives from across the organisation, including front-line colleagues and managers. The idea of the forums is to determine whether a change to the way safety is managed should be implemented; this could be a new policy, a new piece of kit, what best practice actually is. The difference between this sort of forum and the more traditional safety committee is that this forum is empowered to make recommendations to the board as to whether things should be taken forward or not.

As an example, an organisation was aware that their previous approach to safety was very 'top-down' and some of the things they had implemented had not had the benefits they expected as they turned out not to be practical in the 'real world'. To overcome this, the safety professional and the director decided to introduce regional safety forums whose remit was to:

- discuss lost time accidents and key findings;
- review safety audit scores;
- discuss local safety issues;
- identify and recommend best practice;

- raise concerns, suggestions and solutions to safety-related challenges within the region that should be considered as best practice.

The regional forums fed into a national forum made up of representatives to each regional one. They reported to the operational board and were given the remit to:

- review safety performance and trends;
- identify best practice from regions and recommend roll-out across the business to the board;
- define the items for the quarterly branch safety meeting agendas;
- get regional forums to explore and recommend ways to improve specific issues.

It took some time for those attending the forums to build up the trust in their regional manager (and in the case of the national forum, the director) and that it was OK to trust them. Each of the regional forums were tasked to come up with alternatives to safety arrangements that had been in the organisation and were past their use by date, once they had come up with something they liked, they presented it to their peers at the national forum. Provided the national forum signed it off, they would recommend to the operational board that it should be rolled out to the whole organisation.

When the organisation wanted to introduce anything new that impacted on the way front-line colleagues work, a key step in the sign-off process was to take it to the national forum to explain any safety implications and how they would be managed. On one occasion the forum were not happy with some of the arrangements to mitigate the risks arising from the new products to be introduced and refused to sign it off. The project team driving the launch of the new products did not believe that they would not be able to roll it out given what the forum had said. This changed when the operational board said no to it, as they were following the advice of front-line colleagues on the forum.

This example highlights several of the cultural change benefits associated with such a programme:

- People who would not normally get the chance to showcase their talents to their boss had that opportunity and many impressed so much that they were encouraged to go for bigger roles in the organisation and explore different career options.
- The forums turned the normal decision-making process on its head, so that front-line colleagues were given a way to make decisions about things that affect them – within a framework that allowed the organisation to still have sound governance.

- The trust and respect that developed between the front-line colleagues and managers and the business leaders (who chaired the regional forums) as a result of these forums was immense, which helped the organisation achieve other things.
- With the operational board saying no to rolling out a new product because the forum were not happy was an incredibly brave thing to do as it had potential commercial implications, but they did not want to put something in place and lose the commercial benefits through accidents or operational inefficiency it caused.

## Risk assessment working group

Often one of the places where risk assessments fall down is that they tend to be written in isolation by safety professionals, as not only do they see them as 'their job' but sadly so does the organisation. The problem with this, of course, is that the recommendations to control the risks identified can end up being written from a purely health and safety point of view and may not necessarily be reflective of the way the job is actually done or the local challenges that mean the job cannot be done 'as described in the book'.

Clearly, with this more traditional approach the organisation can put a tick in the box that says 'risk assessments completed'. They will fail, though, to get any engagement as well as running the risk that the assessments will not be meaningful. But is engagement necessary with risk assessment? Of course, the answer is yes, because without it the recommendations may not be followed – after all, managers enforcing the rules will only get you so far.

There are many ways to drive engagement with the risk-assessment process and experience shows that a good approach is based on the development of a risk assessment team, as the following example explains. A manufacturing plan wanted to engage more of their front-line workers and managers in the risk assessment to help improve the site's safety performance and culture by moving away from a box-ticking exercise. The site manager and the safety professional set up a risk assessment team made up of members of the workforce from all departments – e.g. production, engineering, distribution (using a similar person specification to one outlined for the safety champion programme discussed above). They were all given training in general risk assessment techniques. Then, in pairs, they were tasked to undertake various assessments taken from a list that the team have developed as part of their training.

The group met every few weeks to review the assessments they had completed and get the recommendations signed off by the safety professional (from a technical accuracy point of view) and also by a site senior manager (one who has the ability to authorise expenditure, can agree safety rules for the location and instruct others to train people in the necessary safe systems of work).

The importance of the safety professional's attendance at these review meetings became clear one day when a member of a risk assessment team presented an assessment following an incident to consider the task of moving two roll cages at the same time. It was explained that the operative was pushing one cage while pulling the other and in order to prevent the following cage striking the back of the operative's ankles when they stopped suddenly safety boots with both steel-toe caps and ankle protection were required. On the face of it, it seemed perfectly reasonable and the site manager was about to instruct someone in the procurement team to source the footwear. At this point the safety professional had to step in and challenged the team's thinking about whether this was the best way to control the risk, or was there another more effective way. The team agreed in the end that the best way to deal with this was not through personal protective equipment, but simply reinforcing an existing rule that cages should be moved one at a time and that they should always be pushed (the only time they should be pulled is when manoeuvring them into a position where they can't be pushed).

The team worked on the principle that if they came across an urgent health and safety situation that needed to be resolved before the next team meeting, they would put the problem and solutions in the form of a risk assessment and present it to the most relevant site manager for their action – in other words, 'Here is the problem, this is what we think *we* need to do about it'. Doing this, everyone found that it is much easier to have an adult conversation about the problem and the solutions.

Of course, there were areas where more specialist risk-assessment training was required (or at least additional knowledge required to undertake the assessments) – for example, when they were looking at chemical safety, manual handling and the like. In these cases assessors from the main team were given additional training to help them develop their specialist skills which can be brought back into the workplace not only in relation to health and safety but also into their 'day job'. For example, an engineer went on a specialist machinery risk-assessment course. As a result, they were not only able to consider the guarding requirements of machines in the factory, but they could also contribute to a working party the site manager was setting up to decide where the machine should be located and how its usage could be optimised.

This approach works very well where there are a number of higher risk activities going on within a single site – in other words, in a production and manufacturing environment. The key process when putting this programme in place is:

- gain site management buy in (based on the saying that many hands make light work);
- identify potential team members;
- determine risk assessments that require undertaking;

- devise and deliver general risk-assessment training;
- assign risk assessments to pairs;
- safety professional maintains high visibility throughout the first few weeks for coaching and guidance;
- risk assessment working group meeting reviews and implements; the programme continues accordingly.

### Front-line manager risk-assessment focus groups

In larger organisations, those which are typically non-production or manufacturing based often use model or generic risk assessments. Such risk assessments are devised by safety professionals centrally and then provided to front-line managers to implement in their workplace and 'tweak' as necessary to make site specific.

As discussed in the risk assessment working group above, it is possible to fall into the 'ivory tower' situation where the people writing the assessments are too far removed from the activities under consideration and thus the solutions that might look good on paper may not translate well in the workplace.

To improve engagement with front-line managers and to understand if there are better ways to do things that suit the operation and the safety professional more, front-line manager risk assessment focus groups are often used and are particularly effective as the following example shows. A multi-site organisation was concerned that their generic risk assessments were not being made site specific by their branch managers. When they did some further investigation, it turned out that the assessments were viewed as being unrealistic and so were discounted by the location managers. Clearly, action was needed. The new safety professional decided to set up a front-line manager risk assessment group to address the problem. They met quarterly with the safety professional to review the new model assessments and were asked to provide feedback along the lines of, 'that would work, that might, that won't' and 'Have you thought of . . .?' When it became clear that those attending were listened to and could positively influence safety within the organisation, they became the early adopters and started to promote the assessments with their peers and, as a result, the uptake improved.

### Safety award programmes

Safety award programmes are implemented on the basis of psychological theory relating to reward and punishment; in a work context this means 'Do well and you'll be rewarded, don't do so well and you won't'. Such programmes do not have to be centred around a reward either. Recognition can be just as powerful – after all, everyone likes being recognised for a job well done. These sorts of programmes simply seek to satisfy this basic human need. For many organisations this is one of the first steps towards a behavioural safety programme and it is easy to implement. There are three fundamental decisions when developing such a programme:

- What behaviours, conditions and performance you intend to recognise or reward.
- How will you measure the achievements?
- What form the reward will take or will it be recognition.

There are many options when it comes to what behaviours, conditions and performance you want to reward. It might be that someone is nominated by their colleagues for being recognised as always working safely – it could be the site that has had the fewest reportable accidents in the year or the location with the best audit score. It does not matter what you decide to recognise or reward, provided it is supportive of the organisation's safety strategy and is a valid measure of safety success. The common trap with safety award programmes is that it can breed a culture of under-reporting and even blame.

This type of programme can be explored further using two examples, the first shows how easy it is to get things wrong, while the second one provides some insights into how to get it right. A large manufacturing organisation operating on a single site ran a programme that was designed to recognise the team that had the fewest lost time accidents (LTAs). The afternoon production team having had three fewer LTAs than their nearest rivals won. The team manager was given a two-week holiday to Cyprus. The logic of this was based on the assumption that the manager must have been managing effectively in order for them to get the result that they did. However, it failed to take into account that it was not just the manager who delivered the result; it was also their team leaders and the front-line employees themselves. Needless to say, the programme's reputation suffered thereafter.

An organisation with a number of call centres ran a programme based on the following criteria:

- All team members must have undertaken their display screen equipment risk assessments within the last year.
- At least one team member per team must have attended every site health and safety committee meeting in the last year.
- It must have received a score of at least 90 per cent on their work area safety inspection (carried out by the Facilities Department) in their last three inspections.

Of the fourteen teams working in the Call Centre, eight achieved the criteria. Each member of the eight teams received 'high street shopping vouchers' to say 'well done!' Everyone was a winner.

## Conclusion

The challenge with actively engaging front-line employees and their line managers with safety is that it is difficult to 'touch' everyone simply because there are so many people in this population. The way to overcome this is to find early adopters and then find ways to work with them to grow your programmes such that they take them to their peers and they then take them to theirs.

It can be incredibly tempting to develop a 'glamorous' behavioural safety programme for front-line workers and managers to drive safety performance and culture change. However, if you have not got the basics of health and safety management in place and working effectively, there is little point. The tools outlined in this chapter enable you to use behavioural safety principles to deal with fundamental health and safety management issues. Arguably, once you get the basics right and working effectively, that is the time to think about a bigger programme.

The tools explored in this chapter can, of course, be moulded to suit each organisation's needs. However, the underlying principle is that you must find ways to get this population of the organisation involved in setting the direction and making the decisions if you are to really deliver outstanding safety performance and lasting beneficial cultural change.

# Part 5

# Keeping going

All the things we have discussed so far – building on your qualifications, understanding the climate we operate in, the way we get things done, thoughts on tackling inaction towards safety and some tools to help engage people at all levels of the business – help to make you become a really effective safety professional. But we haven't talked much about you, how you keep moving forward in your career and development as well as coping with, what can be, an unforgiving role.

Right at the outset we said that getting your safety qualification is only the start of the journey to become a really effective safety professional. We now need to talk about your development. If you do not develop, you become less effective and less sharp. The worst mindset to have – and some people do have this – is 'I have my qualifications (or I've been doing this for years) there is nothing I don't know, I don't need development'. In that one statement alone it proves you really do need some development.

Developing a really useful personal development plan is much more than just thinking about things in the short and medium term. It absolutely should be linked to your much longer term career aspirations. But how do you do that, so that you can achieve your potential? Experience shows that many safety professionals become disillusioned with their role and, make no mistake about it, being a safety professional can be a thankless job. It can also be hugely satisfying but it is tough at whatever level you are as a professional or hope to aspire to become. Even in the most engaged organisations, there will at times be people who do not want to do something, are not interested in safety or what you have to say. You might find yourself under-resourced for what you need to deliver and possibly the most challenging aspect is that real lasting change takes a long time, so if you are the type of person who likes instant satisfaction or to hang your keys up at the end of the day and not think about work until the next day, then you need ways to cope with not getting this all the time. The next two chapters will look to help tackle these two big areas.

# Chapter 22

# Developing your career in safety

Often people tend to give little thought about their own personal development other than what course they would like to go on next, and deciding on their next job also tends to be done with little overall consideration as to whether it will help them get where they want their career to take them. It is easy to see why this happens; we have very few tools to help us put together a really good personal development plan based on you really thinking about both the near and longer term, both of which have very different development needs. Another difficult challenge that we seem less than equipped to face is thinking about how you can achieve your potential and move your way through the world of safety such that you achieve your career ambitions.

In this chapter we will explore the typical requirements of each kind of safety role before going on to understand what sort of development and experiences you need in order to fulfil them successfully – in other words, how to increase your chances of getting those sorts of jobs and being good at them.

## Understanding safety roles

Understanding what each of the typical safety roles does is the first step in putting together a sound development plan either to make you more effective in your current role or to help you achieve your next one. To explore this, we will use a much respected model known as the leadership pipeline (1) which is widely used by organisational development professionals as well as those in HR to establish a pipeline of future leaders from within the organisation, from entry level front-line team leader type roles to chief executives.

The model outlines that there are six leadership levels and moving between them form critical events in a leader's life, as doing so successfully requires different skills. Understanding these can provide you with the insight you need for your development and future career aspirations. Of the six levels, arguably only four really apply to a safety professional's career:

- **Managing self**    This is where you work on your own as a team member developing your technical and professional skills.
- **Managing others**    Here you make the first move into leadership and become a first-time manager of people. The critical thing to learn here is that you do not need to be the best technical expert in order to be effective as a manager of people; the key is to be a good people manager.
- **Managing managers**    In this part of the leadership pipeline, to become effective in the role, you need to become removed from the day-to-day technical aspects of what the team do and focus more on helping your direct reports (the managers of the technical experts) manage effectively through monitoring their team's performance and developing their people management skills.
- **Function manager**    Typically at this level you start to become responsible for some areas that you have little, other than a passing, knowledge of. And as the layers between you and the front line increase, you need to develop new ways to make sure your message, whatever that might be, gets through to them. You also now start to think far less about the here and now, focusing instead on the future strategy and direction of the function.

While not an exact map, Table 22.1 shows the typical safety roles to each of the levels in the leadership pipeline and outlines the sort of work each undertakes.

The first thing you will note in the table is that the safety roles do not generally map that cleanly over to the 'leadership pipeline' levels. This is because often safety functions tend not to have that many layers to warrant the clear-cut levels that the model suggests. However, if you think about other functions in your organisation, the chances are you can see them in use. Indeed, the same thinking applies to 'head of' and director roles in safety as it tends only to be either the high risk or largest organisations that have such positions. There are other twists as well, such as it is common for organisations to have health and safety managers who do not manage anyone. In such cases the job title is more about the culture of the organisation rather than the role itself.

## Your personal development plan

In creating an effective personal development plan you should think about two things:

- What do I need to do to be more effective in my current role?
- What do I need to do to get my next position?

*Table 22.1* Leadership pipeline levels mapped to typical safety roles

| Leadership level | Typical safety role | Typical 'tasks' the role performs |
|---|---|---|
| Managing self | Health and safety co-ordinator<br>Health and safety adviser<br>Fire safety adviser | Administration<br>Health and safety committees<br>Risk assessments, safe systems of work<br>Workplace inspections and audits<br>Incident investigations<br>Delivering training<br>Dealing with enforcement activity<br>Ad hoc support to front-line managers |
| Managing others | Health and safety manager | Review team's work and performance<br>Develop the team<br>Presentations to business teams<br>Manage and deliver business-wide safety improvement programmes<br>Resolve problems that are escalated by your team<br>Develop safety capability of operational managers<br>Identify changes to legislation and best practice that might affect the organisation and implement ways to adapt |
| Manager of managers | Senior health and safety manager<br>Head of health and safety | Develop your managers<br>Inspire and motivate your wider team<br>Agree the organisation's safety strategy and drive its delivery<br>Resolve safety queries raised by the board<br>Report to the board on safety performance, developments and challenges<br>Identify changes to legislation and best practice that might affect the organisation and develop ways to adapt<br>Develop the safety capability of senior managers in the organisation<br>Spend more time understanding how the business works in order to provide partnering type support<br>Interactions with customers, suppliers and other parts of the organisation (if you are in a group) become more as you become the organisation's figurehead for safety<br>Get more involved in areas outside your specialist area and do the above for other non-safety teams that report to you<br>Spend more time networking with peers across other organisations and sectors |

*continued . . .*

*Table 22.1* Continued

| Leadership level | Typical safety role | Typical 'tasks' the role performs |
|---|---|---|
| Function manager | Head of health and safety | The same as manager of managers as well as: |
| | Health and safety director | Develop fellow board and senior leaders' safety capability |
| | | Contribute to wider business decisions, strategy development and execution as part of the board |
| | | Provide oversight and assurance of business safety risks |
| | | Where applicable, lead on the relationship with the non-executive directors, especially where you have a plc safety committee |

In a typical three- to four-year job cycle (the time during which you are in a role), the first eighteen months to two years should be about answering the first of the two questions while the rest of the time should be about developing into your next job.

The leadership pipeline model discussed above is now particularly relevant, as you can see that the further along the safety career ladder you go, the less it is about being able to do the technical aspects of health and safety; as you become increasingly removed from such things, so the type of development you need changes as well.

At the lower end of the spectrum there is still a reliance on technical qualifications and continuing professional development, but you must also be able to put what you have learnt into practice effectively. Nobody wants someone with 'all the qualifications in the world' if they cannot use them to benefit the organisation.

Moving from being in an adviser's position to a first-line manager is quite a leap and to get yourself in the right place to do so, you need to gain some experience of managing people. On first glance, this can seem quite difficult because, as we have already discussed, safety structures tend to be small. However, if you can influence people who do not work for you to do things, then most future employers would be impressed by this. As experience shows, this is quite often one of the most difficult leadership challenges. Fortunately, as safety professionals we have a great number of different people to influence who we indirectly manage – for example, health and safety representatives or risk assessment working groups. Getting on a first-line management development programme, such as that which most organisations run and then putting what you have learnt into practice with safety representatives, is a good way to develop that critical experience.

Moving beyond a health and safety manager's role does become increasingly harder, as the number of positions in this part of the safety profession become fewer. This should not deter you, more make you think about the type of development you need to get the experience necessary to see if you like that sort of role and, to demonstrate you can effectively operate at that level. Many organisations have senior manager and leadership programmes that focus on the softer skills necessary to excel in these positions. Where possible it is time well spent getting on such programmes. Even if there is not a head of or director role in the organisation you work in, the topics you will cover on the programme will most likely help you become a really effective professional.

### Be realistic

There does come a point with this, though, where you need to face some harsh realities. Experience shows that many people are not fully aware of their real abilities and so at times can find themselves in roles that they struggle with, ultimately not performing in and definitely not enjoying. It is important to be realistic in your career aspirations, take time to find out if you are going to like the job you are aiming for, map out how you are going to get there, but remember there is a lot to be said for having a job you can do effectively in an organisation you like working in.

## What to develop

It is worth remembering that every role in the safety profession, to some extent, will require the management and leadership skills discussed in this book, and therefore these topics should be factored into your development plan somewhere. An obvious starting point to decide what you need to develop is what you think you need to do and while this is a good starter for ten, do not rely solely on this. Experience shows that to define a really strong personal development plan you should seek out other people's views on your development needs as they can often see things you cannot.

There are a number of ways you can do this, but by far the easiest is to set up a quick 360 feedback exercise using an online survey tool, many of which are freely available on the Internet. Ask people to respond on a simple rating scale to questions such as 'How good am I at making a robust business case for safety improvements?', 'Do I tend to use the same influencing techniques?' The insight this will give you is priceless.

The next challenge is to think about how you might fill in your development gaps and there are many ways to do it. Your line manager should be able to help as well as your organisation's HR team, but there are things that you can drive as well.

### Gain wide experience

How many times do you see a job advertisement that has in it somewhere 'Candidates must have specific industry sector experience'? In truth, all too often, yet the question the organisation needs to be asking is 'Are they missing something by having that view?' A fundamental reason many people decide to join the profession in the first place is that their underpinning knowledge and base qualification are applicable to all sectors which therefore means you can go off and see a lot of industry and never get bored.

Many safety professionals move from sector to sector every three or four years and each time they make massive safety improvements at the organisations they work for. You can achieve this, in part, because you are new – new to the firm, new to the industry, and so are able to ask the 'why' question more freely than someone who had been in the sector for years or had been with a competitor. Asking why helps you not only to understand what is going on, but also whether things need changing. If the answer you get is sound, then the practice probably really was sound. If the response was anything else, then it might warrant the stone lifting up and having a good look at what lurks beneath. Accept that for some roles, in some organisations, specific experience is necessary, but if you want to have a varied career and hope to take on bigger roles in the future, moving between organisations and sectors is invaluable.

### Secondments

For some people they will not want to leave the organisation they work for, for very valid reasons, but you should not let this stop you from going out and seeing what lessons you can learn from a different organisation both for your own development and also that of your organisation. To help illustrate how this can be achieved, the following example can be used. A safety manager, working for a large service provider, had worked for the firm for over 20 years, having worked in many operational roles they got into safety six years ago. They have many years' 'service' in the pension scheme, so moving to another organisation was far less attractive to them, but they still wanted to see safety in the 'outside world'. They approached their line manager and, in a personal development planning conversation, suggested that they would like to see what other firms do and see what approaches can be brought back to the organisation to help drive safety forward.

After some discussion, the safety manager and their boss agreed to explore the opportunity of them undertaking a secondment at one of their major suppliers. Two months later the safety manager started with the supplier and undertook a three-month placement with them. The supplier gave them a specific project to go and explore, and got them involved in a number of day-to-day activities that helped them see the challenges they faced in managing health and safety.

In the debriefing session after the secondment, the safety manager explained that they were surprised to find just how similar the health and safety challenges both organisations faced were, things like communicating messages to front-line staff that are geographically spread across the country, how to make health and safety interesting and not seen as 'bottom of the list' job. They had also found some new ways to tackle these issues for their own firm and shared some of the things that they had done previously with the supplier.

As a result, each organisation had a fresh angle to come at their issues with and both organisations became a lot closer through this collaboration. A key learning point from this example to help build your confidence to move into different sectors is that we should not think that the organisation you work for is unique in the health and safety challenges it faces. The climate, the organisational tone and the people might be different, but the overarching themes are likely to be the same. We should always try to find ways to break out of the silo and expand our horizons. A secondment in another firm is a great way to get that experience knowing you are going back to your day job at the end of it.

## Conclusion

People often disassociate their personal development plans from their career aspirations or, if they don't, they fail to think about things in the longer term. Being more strategic, in essence using the same sort of approach as we discussed in Chapter 3 about developing a safety strategy, pays dividends to make sure not only you become as effective in your current role as you can but also help you get your next position, and the next, and the next.

If you think that personal development is all about going on courses and getting qualifications, you are limiting yourself. The clue is in the title: personal development. It is yours to be creative in how you meet your development needs. While you will be aware of some of your areas for development, there will be others you are blind to; getting feedback from others is not only powerful but also a very mature approach.

## Reference

1   Drotter, S.J. and Charan, R. *Building Leaders at Every Level: A Leadership Pipeline*. Toronto, Ontario (2001).

# Chapter 23

# Maintaining your drive

Being a safety professional is an important and worthwhile job. You can help improve operational efficiency and sales, improve customer service and relationships as well as employee engagement, reduce cost and help more people to go home in the same condition that they came to work in. However, it is no bed of roses. You will have bad days, but the good days should out-number them.

If you judge how effective you are as a safety professional purely in terms of your organisation's safety performance and culture, you could become quite down-hearted as it often takes time for safety improvements to wash through into the numbers. This also means that the level of job satisfaction you get as a result is lower than you might have in other roles. The important thing is to find the right way of coping with this, because as we saw in Chapter 3 if you have a strategy based on things from a strong situational analysis, the numbers will come good.

In this chapter, then, we will look to address these two final key areas.

## Building up resilience

Resilience refers to our ability to cope with situations such that we 'bounce back' from setbacks or keep going with determination despite resistance. Arguably, this is a necessary requirement for all jobs and life in general, but it is an area at times that experience shows where safety professionals can lack some tools to help them.

### Know why you are in safety

Really understanding why you are in safety in the first place is often the key to being resilient in the role, as this is what you fall back on when things get really tough. It is the reason why you swing your legs out of the bed early on a cold winter's morning to go to work. What is it that drives you?

### Decide what is worth fighting for

We have already talked about working out which fights are worth having and which ones you might be better leaving for now. This idea of accepting that some things you cannot change whereas others you can is a fundamental concept in building up your resilience. Once you can do so things become easier.

### Don't become a catatrophiser

A catatrophiser is someone who, in certain circumstances, can leap to the worst possible outcome of a situation even if in the calm light of day you realise this is not likely. We all catastrophise. It is normal, especially when things start going wrong or people are blocking you. Recognising when you start to feel like this is important and trying to stop it even if it is just vocalising what you are thinking to make you hear yourself being 'silly' is helpful.

### Go back to your strategy

One of the things we said that is very positive about having a safety strategy is that it is something you can reference back to when people start questioning what you are doing and why. Revisit your strategy to make sure what you are doing is in line with that, just checking can make a difference. This helps build confidence in your own abilities: if it is the right thing to do, keep doing it until your boss tells you to stop.

### Step back and see the wood

The saying 'Can't see the wood for the trees' rings true here. Often when you are the in detail you cannot always see the progress you have made. Take stock regularly to review where you are. It will have a massive effect on your resilience and motivation to carry on.

## Developing coping strategies

A coping strategy sounds fancy, but all it is about is finding ways for you to deal with situations so that they do not impact on your well-being, and is integral to developing resilience. It is tempting to say that there are no right or wrong ways to cope with things, but that simply is not true. For example, you could cope with bad days by coming home and drinking a bottle of wine, which clearly can lead to problems.

The important thing, then, is to find coping strategies that do not harm your general health but help you deal with the challenges of being a safety professional. There are many ways to do this and you need to find ways that work for you. To get you thinking along the right lines we will briefly explore ten coping strategies that might help you.

### Don't work in evenings

There is a temptation when you have lots to do to work in the evenings when you come home. Sometimes you will have to do this. However, try to make this the exception not the rule. The same is true for working at weekends. Look to improve your time management (see Chapter 14) so that you can manage your time effectively. You need time to switch off from work. If you do not, you will become less effective when you are there. The number of mistakes you make in your work that you do in the evenings will increase compared to when you are at work and you will struggle to get quality sleep. It is worth remembering that the chances are you will not be thought any more for it anyway.

### Don't let work invade your home life

The worst thing that can happen is that you have had a bad day at work and when you go home you take it out on your partner or the children by, for example, being irritable with them or rushing through your children's bedtime story. Your home should be the place where you can go, shut the door on work and unwind. Experience shows that this is much easier said than done, but try. If when you return home your children or the dog goes mad to see you, spend some time with them. Generally, they will have had enough after a few minutes, allowing you to do what you need to do to 'finish' work.

### Check your emails before you get home

It does not matter what time you leave work. The chances are, thanks to technology, you will have some emails in your inbox waiting for you. The number of them will be much higher if you have been out on a site visit and been travelling for an hour or two. You have three choices: either leave them until the morning, or read them at home, or park before you get home and read. Take whichever option you need to do to help you compartmentalise things – in other words, how you can mentally put work in a box and shut the lid until the next day.

### Switch off your phone

Years ago all mobile phones allowed you to do was make voice calls and send texts. Nowadays, things have changed greatly. For all the benefits this brings, helping you switch off from work is not one of them as it is all too easy to check your emails or send one. Unless you need to keep the phone on or with you, turn it off or put it out of sight. Experience shows that this really helps you switch off and recharge yourself ready for the next day's challenges.

### Getting ready for holidays

Many a great holiday has been ruined by people still thinking about work or having to work on the first few days of their break. Given how much holidays cost and that we never seem to have enough of them, this is not good. If you struggle with this sort of thing, try to find a way round it. You could make a list of all the things you need to do before you go away and then have a day working from home doing just that, or, if it is more that you mentally struggle to switch off, book an extra day's leave at the start of your holiday.

### Exercise

One of the best ways to help you unwind is to exercise; this of course helps other aspects of your general health too. You do not necessarily have to join a gym – any form of exercise or playing sports will help.

### Get a dog

You do not need to get a dog for this, as obviously having a pet is a very serious thing, but dogs are great to help you exercise through regular walks. They are also great listeners and do not answer back. You can tell them your problems and frustrations and they do not judge you. The idea of talking about your frustrations is really helpful too, as getting things out in the open helps you make more sense of them, which is the first step in coming up with a plan to deal with them; in fact, it is the same principle as we discussed in Chapter 11 on problem solving.

### Play music loud

A great way many people find helps them to unwind, switch off and even get rid of some tension is to play music loud on the way home in their car. While making sure you do not go deaf and continue to pay attention to the road and other road users, experience shows this really works.

### Wash your car

If you need instant satisfaction and you sometimes struggle to get it at work, find something out of work that provides you with this to compensate for it. It could be something as simple as washing your car. When you start it is dirty, the wheels have a week's worth of brake dust on them. When you finish it is all shiny and in showroom condition – satisfaction achieved.

### Out-of-work activities

Do not just limit yourself to going to the gym to help you relieve tension. There are many other out-of-work activities that are just as good and often thinking about something or somebody else rather than your situation is really helpful. For example, there are numerous voluntary organisations that are crying out for help.

## Conclusion

Safety professionals play a really important role in organisational life and society. At times we are unsung heroes and at others we get attention for the wrong reasons. Sometimes we have to deal with situations and people who upset us, frustrate us and even disappoint us, but it is all part of the course. Find ways to become more resilient; the more resilient you are the more effective you become. Finding ways to help you unwind and get some quality down time, as well as looking after your health and enjoying your family life will help you even more.

# Index